3rd International Conference on Automation and Mechatronics Engineering (ICAME 2019)

IOP Conference Series: Materials Science and Engineering
Volume 812

Phuket, Thailand
25 - 27 October 2019

ISBN: 978-1-7138-2674-3
ISSN: 1757-8981

Printed from e-media with permission by:

Curran Associates, Inc.
57 Morehouse Lane
Red Hook, NY 12571

Some format issues inherent in the e-media version may also appear in this print version.

Institute of Physics
Dirac House, Temple Back
Bristol BS1 6BE UK

Phone: 44 1 17 929 7481
Fax: 44 1 17 920 0979

techtracking@iop.org

Additional copies of this publication are available from:

Curran Associates, Inc.
57 Morehouse Lane
Red Hook, NY 12571 USA
Phone: 845-758-0400
Fax: 845-758-2633
Email: curran@proceedings.com
Web: www.proceedings.com

TABLE OF CONTENTS

ELECTRONIC AND ELECTRICAL ENGINEERING

CIVIL AND CONSTRUCTION ENGINEERING

IOP Conf. Series: Materials Science and Engineering **812** (2020) 011001 doi:10.1088/1757-899X/812/1/011001

Preface

Automation and Mechatronics Engineering is a key ingredient of current advanced technologies supporting modern society. The tremendous rapid growth of these fields makes them highly interdisciplinary and inevitably demands their mutual interactions. Towards this end, the **2019 3rd International Conference on Automation and Mechatronics Engineering (ICAME 2019)** was organized in Phuket, Thailand during October 25-27 2019.

ICAME 2019 was the conference for Scientists, Researchers, Engineers, Industry Leaders, and Educators to gather at a single platform for discussing the most recent innovations, trends, concerns as well as practical challenges encountered and streamline solutions in the field of Automation and Mechatronics Engineering. The platform provided an opportunity for them to address the latest original results in Automation and Mechatronics Engineering, including both theoretical advances and practical implementations, that are gaining popularity in industries and our daily lives. Furthermore, it also provided an opportunity to highlight recent developments and to identify emerging and future areas of growth in Automation and Mechatronics Engineering.

This proceeding consists of 17 technical papers that are divided into two sub-disciplines namely, Electronic and electrical engineering and Civil and construction engineering. These papers were selected from 30 original contributions, and all papers published in this volume have been peer-reviewed through processes administered by the proceedings Editors.

I wish to extend my gratitude to the keynote speakers, who have accepted our invitation to present their share of knowledge, invited speakers and authors of selected papers for their outstanding contributions. I am also grateful to members of the organizing committee, anonymous reviewers and volunteers for their great efforts.

Prof. Ir. Dr Andrew Ragai Henry Rigit

Organizing Chairman, **ICAME 2019**

Conference Committees

International Advisory Committee
Derong Liu, FIEEE, University of Illinois at Chicago, USA / Guangdong University of Technology, China

General Chairs
Youmin Zhang, Concordia University, Canada
Andrew Ragai Henry Rigit, Universiti Malaysia Sarawak, Malaysia

International Program Chair
Christos Spitas, Nazarbayev University, Kazakhstan

Technical Committees
Semuil Tjiharjadi, Maranatha Christian University, Indonesia
Fasai Wiwatwongwana, Pathumwan Institute of Technology, Thailand
Marina Nikolaevna Dubyago, Southern Federal University, Russia
Mohammed Chadli, University of Picardie Jules Verne, France
Piotr Kulczycki, Polish Academy of Sciences, Poland
Ashraf Darwish, Helwan University, Egypt
José Reinaldo Silva, University of São Paulo, Brazil
Chia-Hung Yeh, National Sun Yat-sen University, Taiwan
Kumar K. Tamma, University of Minnesota, US
Jan Awrejcewicz, Lodz University of Technology, Poland
Haiduke Sarafian, The Pennsylvania State University, USA
Qi Zhou, Huazhong University of Science & Technology, China
Jianbin Qiu, Harbin Institute of Technology, China
Ray-Hwa Wong, Hwa Hsia University of Technology, Taiwan
Juan Manuel Corchado Rodríguez, University of Salamanca, Spain
Yas Alsultanny, Arabian Gulf University, Bahrain
Muhamad Mat Noor,University Malaysia Pahang, Malaysia
Mohamed Arezki MELLAL, M"Hamed Bougara University (UMBB), Algeria
Ashraf Darwish, Helwan University, Egypt
MANJUNATH SHETTAR, Manipal Institute of Technology, India
Paramveer Singh, Lovely Professional University, India
Shivaprakash Y.M, Manipal Institute of Technology, India

IOP Conf. Series: Materials Science and Engineering **812** (2020) 011003 doi:10.1088/1757-899X/812/1/011003

First name	Surname	Institution	E-mail address
Marina N.	Dubyago	Southern Federal University, Russia	w_m88@mail.ru
Maitai	Dahlan	Chulalongkorn University , Thailand	maitai8town@gmail.com
Weibin	Xu	Xinjiang University, China	767910211@qq.com
Z. Q.	Zhou	State Grid Zhejiang Electric Power Research Institute, Hangzhou, China	azzq640105@163.com
E.V.	Martyanov	Researcher, JSC RII «SPECTRUM», Moscow, Russian Federation	EugenioMartino@mail.ru
Junlong	Duan	Department of Engineering and Design, University of Sussex, Brighton BN1 9QT, United Kingdom	J.Duan@sussex.ac.uk
Zuquan	Liang	State Grid Sichuan Comprehensive Energy Service Co.Ltd, China	Liangzuquan@126.com
V. V.	Koryanov	Bauman Moscow State Technical University, 5, b1, 2-nd Baumanskaya Street, Moscow, Russian Federation, 105005	avkoryanov@mail.ru
A. A.	Banishev	Institute on Laser and Information Technologies – Branch of Federal Scientific Research Centre "Crystallography and Photonics" of Russian Academy of Sciences, Svyatoozerskaya Str.1, Shatura 140700, Moscow Region, Russia	banishev@mail.ru
Wei	Guo	Institute of Applied Physics, Jiangxi Academy of Sciences, Nanchang 330096, China	guowei66@sohu.com
Hui	Jiang	Zhongwei National Engineering Research Center For Coking Technology Co., Ltd. , Anshan, Liaoning, 114044, China	asrnls2008@126.com
Dong Won	Jung	Jeju National University, Jeju, South Korea	jdwcheju@jejunu.ac.kr
S	Kumar	BIT Sindri, Dhanbad-828123, India	skumarnitj@gmail.com
A. S.	Sati	Department of Civil and Environmental Eng., University of Sharjah, Sharjah, United Arab Emirates	U17105765@sharjah.ac.ae
K	Pareek	Nanhua University, Chiayi County, Taiwan	10774107@nhu.edu.tw
S	Kumar	BIT Sindri, Dhanbad-828123, India	skumarnitj@gmail.com
Chi	Tang	National Pingtung University of Science and Technology, Pingtung, Taiwan	tangchi@mail.npust.edu.tw

Peer review statement

All papers published in this volume of *IOP Conference Series: Materials Science and Engineering* have been peer reviewed through processes administered by the proceedings Editors. Reviews were conducted by expert referees to the professional and scientific standards expected of a proceedings journal published by IOP Publishing.

IOP Conf. Series: Materials Science and Engineering **812** (2020) 012001 doi:10.1088/1757-899X/812/1/012001

A traction controller of a two-wheel mobile robot with feedforward compensator

M Dahlan[1], A Sutapun[1] and V Sangveraphunsiri[1]

[1]Chulalongkorn University, Department of Mechanical Engineering, 254 Phayathai, Wang Mai, Pathum Wan, Bangkok, 10330, Thailand

Email: maitai8town@gmail.com

Abstract. A simple traction controller was derived from a model following control (MFC) using bilinear transformation. The controller was simplified by adjusting some dynamic parameters. The experiments were conducted to evaluate the validity of the controller by comparing with the classical PD controller. In the experiments, while the traction controller can prevent slippage, tracking control performance needed to be improved. The feedforward compensation was added to reduce the traction error. The experimentation using the simple controller with feedforward shows that the position error has been reduced while maintaining the traction force between floor and wheels.

1. Introduction

With technology becoming more advanced every single day, robots are more affordable than ever, thanks to the lower price and higher efficiency of computational powers and batterie. Wheeled mobile robots have a wide variety of application from house vacuums cleaning to warehouse stocking management. This project's main focus is to develop a hospital mobile assist robot for caring and rehabilitation services for the elderly and post-stroke patients.

For safety in the hospital, the mobile robot must have maneuverability advantage such as obstacle detection, dynamic mapping, obstacle avoidance system, traction control. There is much research in those domains [1,2]. The focus on this paper, traction control is essential in the safety aspect as the slip can cause many problems, for example, unintentional rotation, less optimum break distance, the error between actual longitudinal motion and senses motion, tire mark, etc.

Traction control and so many advanced control algorithms were researched and developed for electrical vehicles [3-8]. Those control algorithms can be applied in the robotic field to expand the range of the real-world application such as two-wheel mobile robot driven on ice [9].

The traction control algorithm in this research is based on the model following control method (MFC) [10], which is one of many early traction control algorithms. The main advantage of MFC is it used only wheel angular sensor to prevent a slip which is good for mobile robots that have so little space to fit more sensors. Originally, MFC needs three constant parameters to be chosen. In this work, simplify version of MFC was researched and used. The simplify version of MFC only needs one gain parameter to implement. The result compared between 3 controllers including, the tradition PD controller, simplified MFC controller and simplified MFC controller with feedforward compensation. The latter was tested to find an improvement to the tracking problem occurred in simplified MFC controller.

This paper has 6 main sections as follows: Section 2 describes the mobile robot used in this research. Section 3 states the controllers used in the experiments. Section 4 shows and compares the result of the experiments. Section 5 and section 6 are result and conclusion respectively.

IOP Conf. Series: Materials Science and Engineering **812** (2020) 012001 doi:10.1088/1757-899X/812/1/012001

2. The two-wheel mobile robot

Fig. 1 shows the two-wheel mobile robot used in this research. Each wheel is driven by a brushless motor with 1:6 gear ratio planetary gear. This robot uses a differentially steered drive configuration with rotating against the center of the robot. The driven wheel connected to the gear is made of milled aluminum coated with rubber. Four caster wheels are installed at the four corners to support the robot. The driven wheels also hang on the spring suspension mechanism to keep the contract between the wheel and ground. The specifications of the robot are shown in Table 1.

Figure 1. The two-wheel mobile robot.

Table 1. The specifications of the mobile robot.

Specifications	
Stall torque	78 Nm
Max. continuous torque	5.784 Nm
Maximum speed	1.60 ms^{-1}
Maximum power	260 W
Wheels diameter	20 cm
Encoder	2500×4 pulse
Dimension	57×50×30 cm
Sampling rate	2000 Hz

3. The control schemes

3.1. Proportional–derivative controller

Fig. 2 shows the block diagram of the PD controller. Reference position (θ_r) and velocity ($\dot{\theta}_r$) are determined by the trajectory planner then the position (θ_m) and velocity ($\dot{\theta}_m$) are fed back to perform closed-loop control with gain K_p and K_d. The wheel velocity calculated by performing derivation and low pass filtered on the position signal.

So, the control law is

$$u = K_p(\theta_r - \theta_m) + K_d(\dot{\theta}_r - \dot{\theta}_m) \tag{1}$$

3.2. Traction controller

Many works have been done in the traction control algorithm, e.g. traction control with slip estimator for a rocker-bogie robot [11], traction control with dynamic tire friction model [12], optimal slip ratio control and model following control [10].

The model following control (MFC) is simple to implement as it does not need the implementation of the tire-floor friction estimator or slip ratio estimator. MFC has also been verified through experiments for simple traction control.

To prevent the slip phenomenon, the modified implementation of the MFC is utilized in this two-wheeled robot. The block diagram of MFC is shown in Fig. 3.

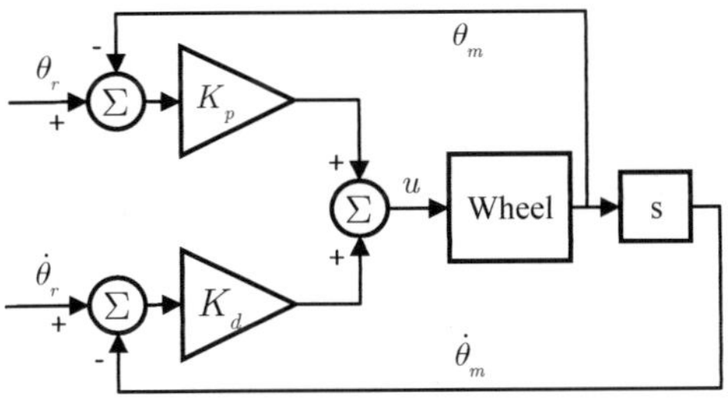

Figure 2. The proportional–derivative controller

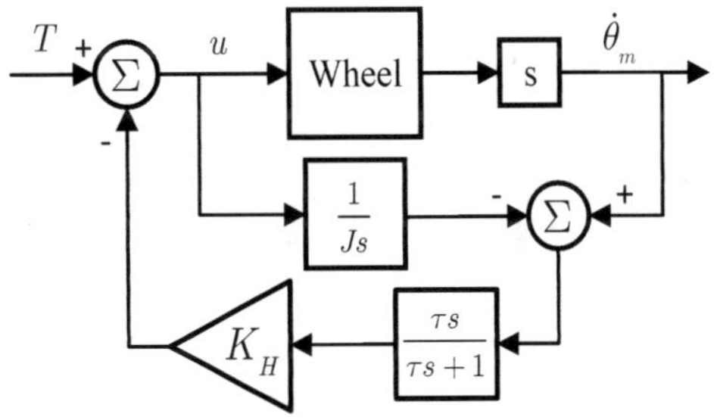

Figure 3. The model following controller

The MFC prevent slip by reducing the output with the signal from a slip estimator. The output from the position and velocity controller (T) is used to estimate the wheel velocity and the slip is estimated by filtering the different of the measured velocity and the estimated velocity. The time constant of the high pass filter (τ), gain K_H, and system inertia (J) need to be chosen for this model. If slip, the sudden change of the measured velocity will be detected, and the slip will be reduced by the controller.

The output from the block diagram is

$$u = \left[T\left(s+\tau^{-1}\right) - K_H\dot{\theta}_m s\right]\left(s+\tau^{-1} - K_H J^{-1}\right)^{-1} \tag{2}$$

For implementation, the bilinear transformation was utilized. The bilinear transformation is a mathematical relationship which is used to convert the complex Laplace domain (s-domain) into z-domain. Using the substitution

$$s = \frac{2}{t}\frac{z-1}{z+1} \tag{3}$$

Where t is the system sampling time and z is the complex function variable. And then

$$u = \frac{T\left[2(z-1)+\tau^{-1}t(z+1)\right] - K_H\dot{\theta}_m 2(z-1)}{\left[2(z-1)+\tau^{-1}t(z+1) - t(z+1)K_H J^{-1}\right]} \tag{4}$$

If the sampling time t is small enough, the part with t can be neglected

$$u = uz^{-1} + T\left(1-z^{-1}\right) - K_H\dot{\theta}_m\left(1-z^{-1}\right) \tag{5}$$

with

$$T = K_p(\theta_r - \theta_m) + K_d(\dot{\theta}_r - \dot{\theta}_m) \tag{6}$$

After the control law is simplified by removing the part with a very small coefficient, the high pass filter time constant and the system inertia are neglected, making the control law dynamically independent.

The output from the position and velocity controller T is used together with the wheel velocity signal $\dot{\theta}_m$, calculated from the derivative of θ_m, to make a slip prevention controller. The gain K_H can be chosen to achieve the desired traction force.

3.3. Traction controller with feedforward compensation

The feedforward compensation helps the tracking problem in a simplified MFC controller scheme. The overall block diagram is shown in Fig. 4.

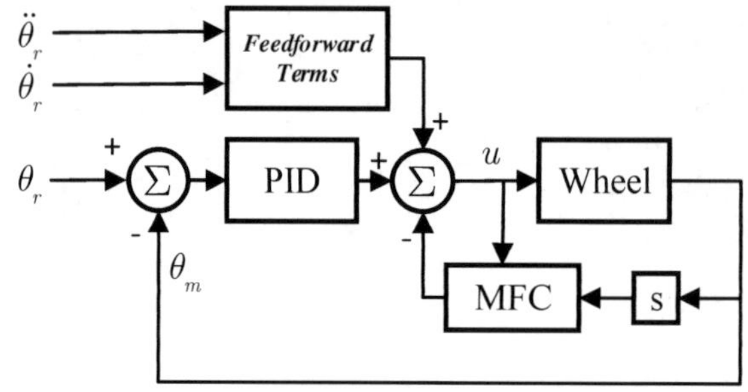

Figure 4. The traction control with feedforward controller

Reference position (θ_r), velocity ($\dot{\theta}_r$) and acceleration ($\ddot{\theta}_r$) are generated by the trajectory planner and use as a feedforward signal combine with the PID controller. The MFC part is the same as the previous control scheme.

4. The experiments

The trajectory used in this paper is generated by using the trapezoidal velocity trajectory profile. The first experiment compares PD controller with simple traction controller using parameters as follows: maximum velocity and acceleration are 1.3 ms^{-1} and 1.5 ms^{-2} respectively. The last experiment compared only simple traction controller and simple traction controller with feedforward compensator using parameters as follows: maximum velocity and acceleration are 1.0 ms^{-1} and 1.5 ms^{-2} respectively. Both trajectories travel the same distance of 2 meters.

The mobile wheels are made of rubber and the floor is made of polished stone. The trajectory of each wheel of the robot is generated with the same parameter, that is, same position, same velocity, and same acceleration at any point in time. Ideally, the robot will move straight if slips did not occur.

5. Result and discussions

5.1. Experiment 1

Fig. 5 shows trajectory data from the experiment including position, velocity, and acceleration. Fig. 6 illustrates the actual motion robot from the starting position at $t = 0\,sec$ to the stop position at $t = 3\,sec$ (right to left). The robot in the Fig. 5a and 6a was driven by the PD controller. The robot in the Fig. 5b and 6b was driven with the traction controller. The robot will move straight if both wheels rotate at the same speed and exert the same amount of traction force. On the contrary, if any of the wheels is slipped, the forces become unbalanced and the generated coupling torque will make the robot rotate unintentionally.

From the acceleration graph in Fig. 5a, the spike of the acceleration at the beginning shows that a wheel is slipped, corresponding with the robot in Fig. 6a. The slip causes the robot to turn and drive off the course. The slip also occurred at the end of the motion, generated by the deacceleration, which make the robot turn again but in the opposite direction.

The traction control law result was evaluated in Figure 4b. The velocity gradually increased without any spike in the acceleration signal. The traction control was able to suppress the slip. From the robot motion in Fig. 6b, the slip did not happen as the robot drive in a straight line and did not have any observable rotation motion both at the start and at the stop.

While this controller can prevent slip phenomenon, the traction error that occurs was very noticeable.

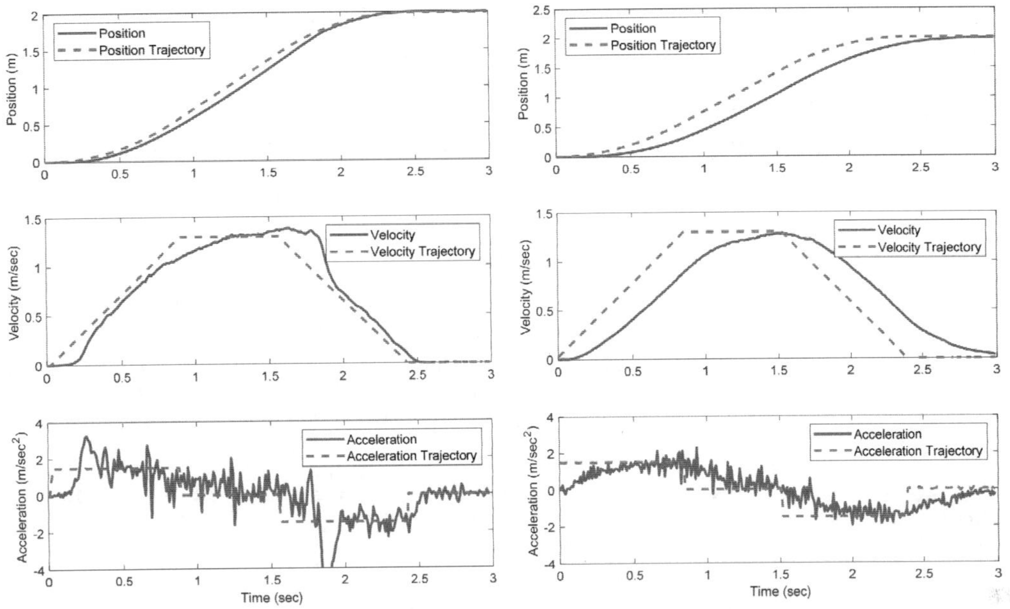

(a) (b)

Figure 5. The trapezoidal velocity trajectory tracking by (a) the position and velocity controller, (b) the traction controller

(a) (b)

Figure 6. The motion of the robot while driven with (a) the position and velocity controller, (b) the traction controller

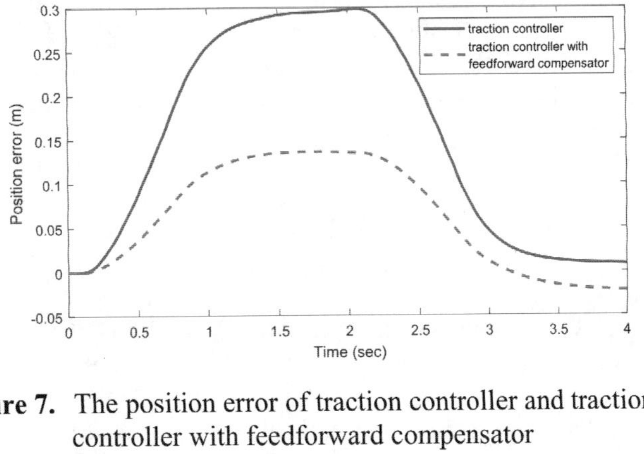

Figure 7. The position error of traction controller and traction controller with feedforward compensator

5.2. Experiment 2

This experiment was comparing the position error between only simple traction controller and simple traction controller with feedforward compensator as shown in Fig. 7.

The result shows that the simple traction controller with feedforward compensator can significantly reduce the traction error. The error can be reduced further by gain tuning in the real environment.

6. Conclusion

The MFC has proven effective on both EV and wheeled mobile robot but still required suitable dynamic parameters for implementation. Nevertheless, in the discrete domain, the MFC controller can be made simpler by adjusting some terms in the control law. This approximation eliminates the dynamic parameter. The remaining terms in the control are related to state variables and gain parameter.

From the experiment, the slip phenomenon can be suppressed with the simplified MFC traction controller. The experiment also shows that the controller can be improved by adding feedforward compensator, making the traction error more subtle.

Future work will involve the estimation of the maximum traction force on any surface-pair using traction control on the wheeled mobile robot. Exploring other sophisticated traction controllers will be also considered.

7. References

[1] J. Borenstein, H.R. Everett, L. Feng, D. Wehe, "Mobile robot positioning: Sensors and techniques," J. Robot. Syst. vol. 14, no. 4, p. 231–249, 1995.

[2] W. Poomarin, R. Chancharoen, V. Sangveraphunsiri, "Automatic docking with obstacle avoidance of differential wheel mobile robot," Int. J. Mech. Eng. Robot. Res., vol. 5, no. 1, p. 11–16, Jan 2016.

[3] K. Fujii, H. Fujimoto, "Traction Control based on Slip Ratio Estimation Without Detecting Vehicle Speed for Electric Vehicle," IEEE PCC'07, p. 688–693, Apr 2007.

[4] G. A. Magallan, C. H. De Angelo, G. O. Garcia, "Maximization of the Traction Forces in a 2WD Electric Vehicle," IEEE Transactions on Vehicular Technology, vol. 60, no. 2, p. 369–380, 2011.

[5] Y. Furukawa, N. Yuhara, S. Sano, H. Takeda, Y. Matsushita, "A Review of Four-Wheel Steering Studies from the Viewpoint of Vehicle Dynamics and Control," Vehicle system dynamics, vol. 18, no. 1-3, p. 151–186, 1989.

[6] E. Ono, Y. Hattori, Y. Muragishi, K. Koibuchi, "Vehicle dynamics integrated control for four-wheel-distributed steering and four-wheel-distributed traction/braking systems," Vehicle system dynamics, vol.44, no. 2, p. 139–151, 2006.

[7] S. Oh, K. Kong, Y. Hori, "Disturbance Attenuation Control for Power-Assist Wheelchair Operation on Slopes," Mechatronics, vol. 24, no. 8, p. 1101–1111, 2014.

[8] K. Nam, S. Oh, H. Fujimoto, Y. Hori, "Estimation of sideslip and roll angles of electric vehicles using lateral tire force sensors through RLS and Kalman filter approaches," IEEE Transactions on Industrial Electronics, vol. 60, p. 988–1000, 2013.

[9] J. H. Choi, S. Oh, "Traction Control for Two-Wheel Driven Mobile Robot Driving on Ice," IEEE International Symposium on Industrial Electronics, vol 2018–June, p. 1075–1080, Jun 2018.

[10] Y. Hori, Y. Toyoda, Y. Tsuruoka, "Traction control of electric vehicle: basic experimental results using the test EV "UOT electric march"," IEEE Trans. Ind. Appl., vol. 34, no. 5, p. 1131–1138, 1998.

[11] M. Thianwiboon, V. Sangveraphunsiri, "Traction Control for a Rocker-Bogie Robot with Wheel-Ground Contact Angle Estimation," Robocup 2005: Robot Soccer World Cup Ix, p. 682–690, 2006.

[12] C. Canudas de Wit, P. Tsiotras, "Dynamic tire friction models for vehicle traction control," Proceedings of the 38th IEEE Conference on Decision and Control, vol. 4, p. 3746–3751, Dec 1999.

IOP Conf. Series: Materials Science and Engineering **812** (2020) 012002 doi:10.1088/1757-899X/812/1/012002

Stabilization of Power Fluctuation of Renewable Energy Based on Hybrid Energy Storage System

Weibin Xu[1], Xinyan Zhang[1,2], Jiajun Zhang[1], Lei Yu[1] and Yang Yu[1]

[1]School of Electrical Engineering, Xinjiang University, 830047 Urumqi, China
[2]Research Center of Renewable Energy Generation and Grid-connected Control Engineering Technology, Ministry of Education, 830047 Urumqi, China

E-mail: 767910211@qq.com

Abstract. As energy shortages become more prominent, renewable energy is widely exploited. Renewable energy sources such as wind energy and solar energy are affected by the non-uniformity and non-steadiness of the natural environment, resulting in fluctuations and randomness of output power. When incorporated into the grid, it will cause fluctuations in grid voltage and frequency, affecting the power quality of the grid. The energy storage system can be used as an energy buffering device to stabilize the grid-connected power to meet the grid-connected standard. A hybrid energy storage system is proposed to stabilize the fluctuation of renewable energy generation, and the energy storage control method and energy distribution method are given. Finally, using the actual data of a wind farm, the hybrid energy storage composed of hydrogen storage energy and supercapacitor energy storage is used to stabilize the wind power fluctuations. The simulation results verify the technical rationality and economic practicability of the proposed method.

1. Restrictions on the development of renewable energy

Developing and utilizing renewable energy is an important way to solve the problem of energy exhaustion. The consumption of renewable energy is one of the key technologies that restrict the development of renewable energy. Wind energy, solar energy and other renewable energy have some shortcomings, such as discontinuity, instability, low density, time, season and climate change. The power quality of power grid decreases when these energy-generating equipment is connected to the grid[1].

Taking wind energy as an example, the output of wind turbines is stochastic and intermittent because the wind power is affected by the change of natural conditions and the wind speed and direction are changing constantly and randomly. The incorporation of wind farms into power grids may lead to serious harmonic pollution and voltage flicker in local power systems. Because of the particularity of electric energy, generation, transmission, distribution and power consumption in power system must be completed at the same time, which requires that the system should always be in a dynamic equilibrium state. Wind has uncontrollable and iradjustable characteristics, which will cause the security and stability of the system. The above factors lead to a large number of abandonment of wind and light. As shown in Table 1, the abandonment of wind power in China's major wind power provinces.

Table 1. Abandonment of wind in major wind power provinces in China.

Province	Power generation capacity(billion kW·h)	Abandoned wind capacity(billion kW·h)	Abandoned wind rate(%)
Inner Mongolia	46.4	12.4	21
Shanxi	13.5	1.4	9
Ningxia	12.9	1.9	13
Gansu	13.6	10.4	43
Xinjiang	22.0	13.7	38
Hebei	21.9	2.2	9
Liaoning	12.9	1.9	13
Heilongjiang	8.8	2.0	19
Jilin	6.7	3.0	29

It can be seen that China's wind abandonment situation is relatively serious. Therefore, how to improve the stability of renewable energy generation system and restrain the power quality problems caused by intermittent renewable energy generation system is an important issue in the grid-connected operation of renewable energy generation system[2-3]. Energy storage technology can store electric energy, release it when needed, ensure the continuous and stable output of electric energy, and improve the power grid's ability to accept renewable energy for power generation [4].

2. Renewable energy and hydrogen storage system

2.1. Energy storage technology

Energy storage technology is divided into physical energy storage, chemical energy storage and thermal energy storage. Physical energy storage includes mechanical energy storage (e.g. pumping energy storage, compressed air energy storage, flywheel energy storage) and electromagnetic energy storage (e.g. supercapacitor and superconducting energy storage); chemical energy storage is based on electrochemical principles (e.g. lead-acid battery, lithium-ion battery, sodium-sulfur battery, etc.); thermal energy storage is to store heat energy in the medium of heat-insulating container to realize thermal energy generation or direct benefit. Use. Fig. 1 is a comparison of the duration of sustainable power supply and storage capacity of commonly used energy storage.

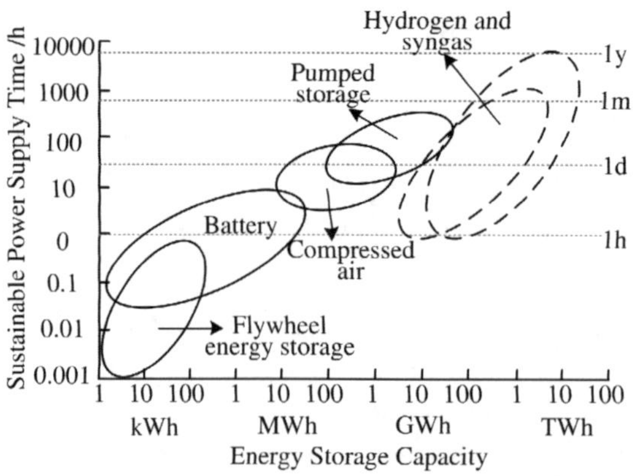

Figure 1. Contrast Chart of Energy Storage Mode.

Figure 1 shows that hydrogen storage has the best comprehensive performance in terms of power supply time and energy storage capacity. The advantages of hydrogen as an energy carrier lie in: (a) high efficiency conversion between hydrogen and electricity can be achieved through electrolytic water and fuel cell technology; (b) compressed hydrogen has high energy density; and (c) hydrogen has the potential to scale up to power grid applications.

2.2. Hydrogen storage system for renewable energy generation

Renewable energy generation hydrogen storage system uses renewable energy (solar energy, wind energy, etc.) to produce hydrogen by electrolyzing water and storing hydrogen. When needed, it generates electricity through fuel cells to supplement electricity. Its basic structure includes: electrolytic water hydrogen production system, hydrogen storage system, fuel cell power generation system, energy management and control system, etc. Its schematic diagram is shown in Figure 2.

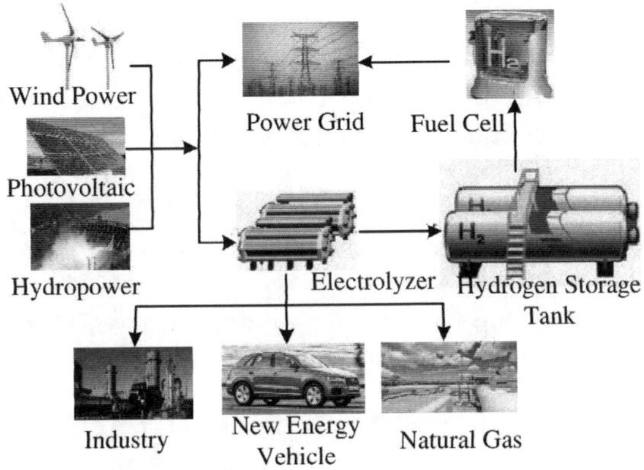

Figure 2. Hydrogen storage system.

2.3. Key technologies in hydrogen energy storage system

The hydrogen energy storage system uses the clean energy power electrolysis technology to obtain hydrogen, stores the hydrogen in the hydrogen storage device, and then uses the fuel cell technology to feed the stored energy back to the power grid, or sends the stored high-purity hydrogen to the hydrogen industry chain. . In order to realize this energy conversion system, the technologies used include: hydrogen production technology, hydrogen storage technology, and hydrogen power generation technology. Further breakthroughs in these key technologies are needed. Therefore, the development of high efficiency, long life, low cost and high performance hydrogen storage technology for power grid is one of the key technologies to promote the application of hydrogen storage in new energy access.

3. Control strategy of hybrid energy storage for suppressing wind power fluctuation

The control strategy of hybrid energy storage to suppress wind power fluctuation is an important means to suppress wind power in order to improve the operation ability of wind power access[5]. The wind power hybrid energy storage power generation system includes wind turbine, hybrid energy storage system and hybrid energy storage power generation system. The structure of wind power hybrid energy storage system is shown in Figure 3.

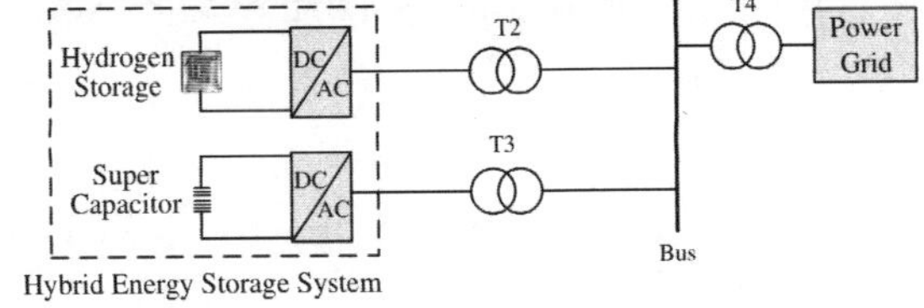

Figure 3. Wind power hybrid energy storage system

When the grid can not absorb wind power, surplus energy can be allocated to hydrogen storage system for energy storage. The system is composed of wind power generation system, hybrid energy storage system, power distribution controller and connecting wires. When wind power is excessive, energy flows from busbar to hybrid energy storage system. When wind power is insufficient, energy flows from hybrid energy storage system to busbar and supplies power grid. Hybrid energy storage system is composed of hydrogen energy storage and supercapacitor energy storage. The response speed of hydrogen energy storage is slow, but its storage capacity is large, and hydrogen energy storage can bear low-frequency power. Supercapacitor has a fast response speed and is suitable for frequent charging and discharging. Supercapacitor bears high frequency power. Energy storage system can play a "peak cutting and valley filling" effect on the output power of wind power. The input power of the system is smoother.

At present, there are many wind power fluctuation mitigation strategies, such as empirical mode decomposition, wavelet decomposition, Kalman filter, first-order filtering, and so on[6-7]. But these suppression strategies have some limitations in real-time online control. Model predictive control (MPC) has the characteristics of early prediction, priority control and good real-time performance. Therefore, this paper adopts model predictive control (MPC) strategy to suppress wind power fluctuation. Model predictive control strategy includes rolling time domain strategy and state space model[8-9].

In actual operation, the relationship between the output power of wind farm and energy storage is as follows:

$$P_g(k+1) = P_{es}(k) + P_w(k) \tag{1}$$

In the formula: $P_g(k)$, $P_{es}(k)$ and $P_w(k)$ are the grid-connected power, energy storage power and original power of wind power at k time respectively. The formula for calculating the state of charge of energy storage at k+1 time is as follows:

$$SOC_{es}(k+1) = SOC_{es}(k) - \frac{T_c P_{es}(k)}{C_{es}} \tag{2}$$

In the formula: T_c is the energy storage control period; C_{es} is the energy storage capacity.

According to the principle of state space model, P_g and SOC_{es} are taken as state variables x_1 and x_2, P_{es} as control variables u, P_w as disturbance input r, P_g and SOC_{es} as output variables y_1 and y_2, and then the state space expression is constructed as follows:

$$\begin{cases} \begin{bmatrix} x_1(k+1) \\ x_2(k+1) \end{bmatrix} = \begin{bmatrix} 0 & 0 \\ 0 & 1 \end{bmatrix} \begin{bmatrix} x_1(k) \\ x_1(k) \end{bmatrix} + \begin{bmatrix} 1 \\ -T_c \end{bmatrix} u(k) + \begin{bmatrix} 1 \\ 0 \end{bmatrix} r(k) \\ \begin{bmatrix} y_1(k) \\ y_2(k) \end{bmatrix} = \begin{bmatrix} 1 & 0 \\ 0 & 1 \end{bmatrix} \begin{bmatrix} x_1(k) \\ x_1(k) \end{bmatrix} \end{cases} \tag{3}$$

According to the idea of rolling time domain strategy, iterative calculation of equation (3) can predict the output of variables in the future $k+M$ (M=1, 2, 3,...), and then use the quadratic programming method in each control. Optimization in the time domain, you can get the corresponding control instructions.

The above analysis is transformed into an optimization model solution, in which the objective function is the minimum energy storage usage J in each optimization cycle, namely:

$$J = \min(\sum_{i=k}^{k+M} P_{es}^2(i)) \tag{4}$$

Accordingly, the constraints of energy storage power, power fluctuation rate gamma of wind power Grid-connected and energy storage charging state should be satisfied, and the constraints are as follows:

$$\begin{cases} P_{es\_\min} \le P_{es}(i) \le P_{es\_\max} \\ \dfrac{P_{g\_\max}(i) - P_{g\_\min}(i)}{P_{rated}} \le \gamma \\ 0 \le SOC_{es}(i) \le 1 \end{cases} \tag{5}$$

In the formula: P_{es_min} and P_{es_max} are the lower and upper limits of P_{es} respectively; P_{g_min} and P_{g_max} are the minimum and maximum values of P_g respectively; P_{rated} is the rated installed capacity of wind power.

4. Case analysis

A day's actual power data of a wind farm is selected as the research object. The sampling time is 1440 minutes, the sampling rate is 1 minute, and the installed capacity is 200 MW. Figure 4 shows the waveform of the actual output power of a wind farm.

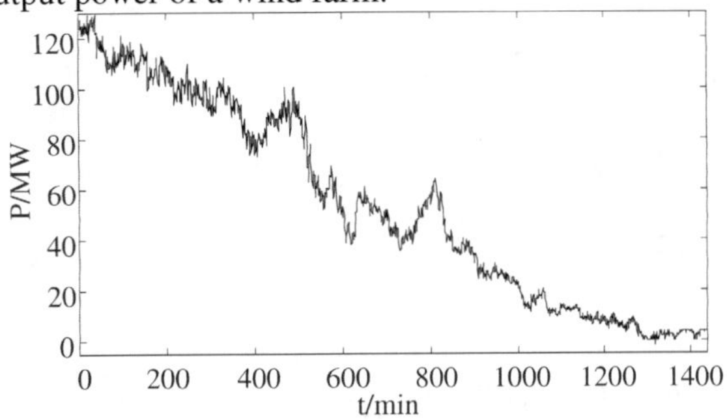

Figure 4. Actual output power of wind power

4.1. Stabilization strategy

According to the constraints of Formula (5), 1 minute fluctuation rate ≤2% and 30 minute fluctuation rate ≤7% are selected as the wind power Grid-connected fluctuation rate constraints. The storage power constraints Pes takes 9MW and the storage *SOC* range is 0-1. The model predictive control algorithm is used to suppress the wind power curve. The wind power before and after the suppression is shown in Figure 5.

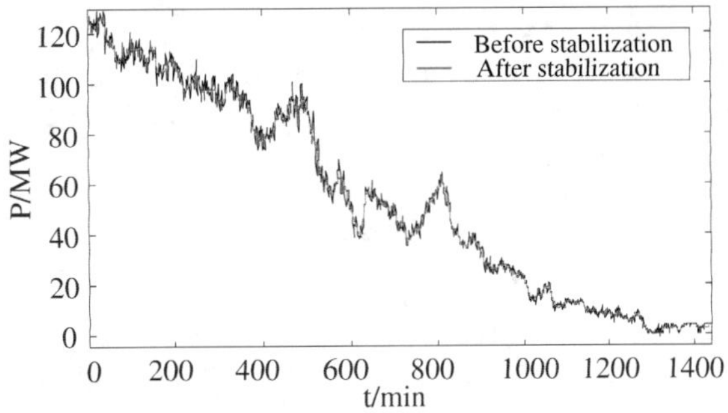

Figure 5. Actual output power of wind power

It can be seen that the wind power fluctuation after smoothing is relatively smooth, and the fluctuation rates of 1 min and 30 min after smoothing are shown in Figs. 6 and 7 respectively.

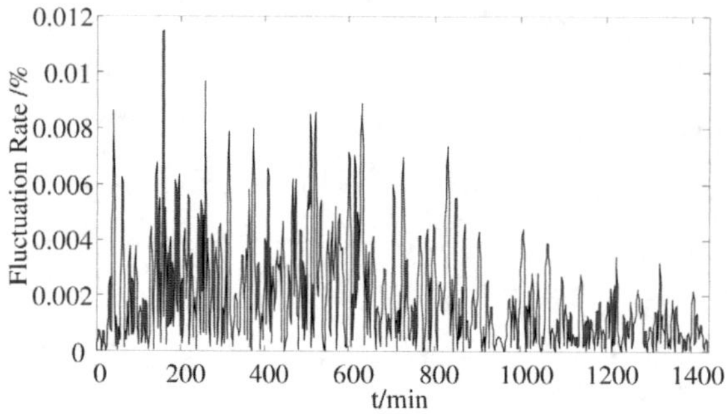

Figure 6. Volatility of 1 minute after stabilization

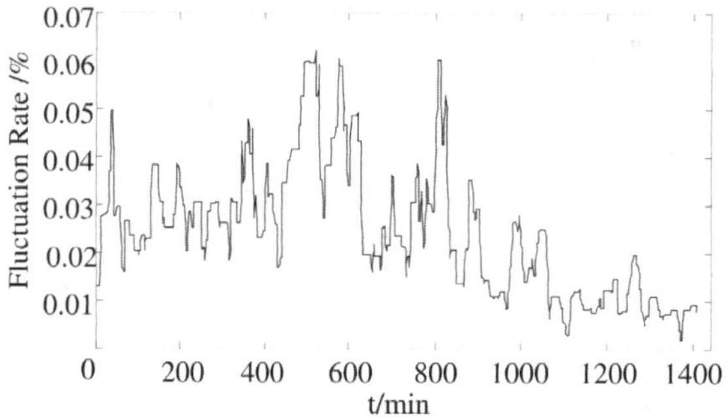

Figure 7. Volatility of 30 minute after stabilization

From Figs. 6 and 7, we can see that the volatility after stabilization decreases greatly. The maximum volatility after stabilization is 1.19% in one minute and 6.17% in 30 minutes. To meet the grid-connected requirements: 1 minute fluctuation rate after stabilization is less than less than or equal to 2%, 30 minute fluctuation rate is less than or equal to 7%.

4.2. Electric power distribution

Considering the large storage capacity and slow charge-discharge response of hydrogen energy storage, hydrogen energy storage is used to undertake fluctuating low-frequency power instructions. The charging and discharging response time of supercapacitors ranges from milliseconds to tens of minutes, which is used to undertake high-frequency power instructions with frequent fluctuations. In order to meet the performance characteristics of hydrogen energy storage and supercapacitor, the energy distribution of hydrogen energy storage is shown in Figs. 8 and 9.

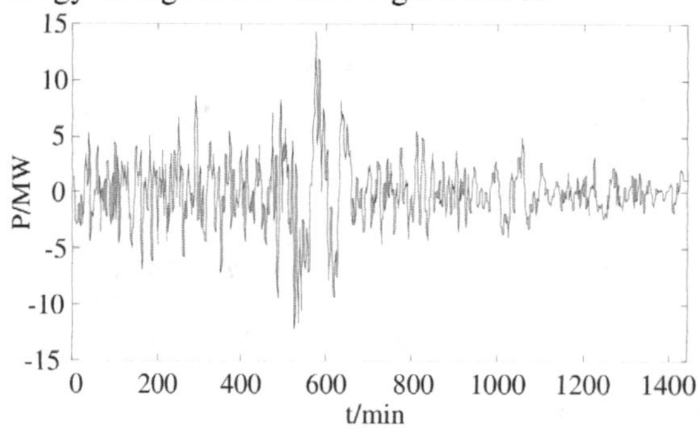

Figure 8. Hydrogen storage power

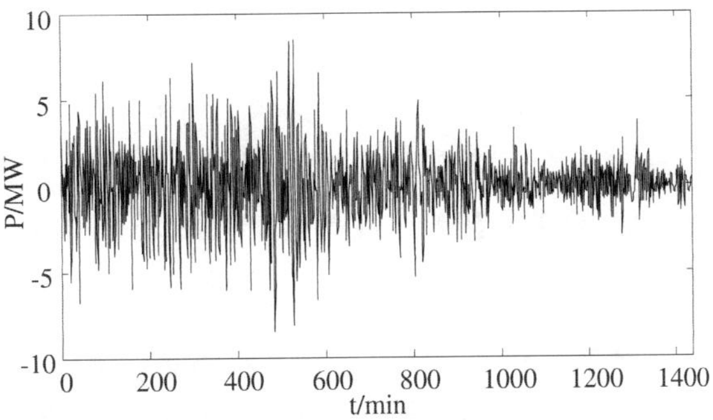

Figure 9. Supercapacitor storage power

5. Summary

The randomness and fluctuation of wind and photovoltaic renewables bring great challenges to the stability and reliability of voltage and frequency of large power grid when wind and photovoltaic are connected to the grid. Energy storage system can be used as energy buffer device, which can be widely used in the field of renewable energy generation control with intermittent output power to enhance the stability and controllability of its output power and improve the power quality to meet the grid-connected standard of wind power. Through the simulation and analysis of hybrid energy storage system, the results demonstrate that the hybrid energy storage system can effectively suppress the random, intermittent and fluctuating characteristics of wind power output, greatly enhance the wind power absorption capacity and the stable operation of wind power, thereby improving the utilization rate of renewable energy generation, and provide technical support for the large-scale development and utilization of renewable energy.

6. References

[1] G. Xu, H. Cheng, Z. Ma, S. Fang and Z. Ma 2017 *Power grid technology* (China) p 3470-3479
[2] A. Basit 2016 *Journal of Modern Power Systems and Clean Energy* (Denamark) p 229-237
[3] F.M. Gonzalez-Longatt 2016 *Journal of Modern Power Systems and Clean Energy* (UK) p 211
[4] T. Shi, B. Zhang and Q. Chao 2016 *Power grid technology* (China) p 477-483
[5] T. Xie, J. Lu and Z. Weng 2016 *Power grid technology* (China) p 2052-2058
[6] Q. Jiang and H. Hong 2013 *IEEE Transactions on Power Systems* (China) p1363-1372
[7] Q. Jiang and H. Wang2013 *IEEE Transactions on Energy Conversion* (China) p 52-61
[8] Y. Sun, X. Li and X. Tang 2017 *High voltage technology* (China) p 181-188
[9] Y. Sun, G. Zhang and X. Tang 2018 *High voltage technology* (China) p 3407-3416

Experimental Studies of the Corrosion Resistance for Q235 and Q235NHin the Outdoor High-voltage Disconnector

Z Q Zhou[1], Y T Zhou[1], J JHe[2], Y Lu[1], Q F Sun[1], H Y Mao[1]

[1]State Grid Zhejiang Electric Power Research Institute, Hangzhou, China
[2]State Grid Zhejiang Zhuji Power Supply Company, Zhuji, China

E-mail: azzq640105@163.com

Abstract. Aiming at the corrosion failure question of the outdoor high-voltage isolation switch, accelerated corrosion testsusing 5% (mass fraction) NaCl salt spray for materials of Q235 and Q235NH were carried out. By means of the controlled trials method, samples were divided into two groups, and each group contains four specimens. After that, neutral-salt-spray corrosion tests with 60h, 100h, 130h, and 160h were conducted. Finally, the rust layer was observed by a microscope, and the corrosion rate was calculated according to the weight loss method. Experimental results indicated that the rust layer in Q235NH was more compact after being corroded. As a result, it provide more protection for the substrate.

1.Introduction

In the power plant and the substation electrical system, the high voltage disconnector is the most widely used device[1]. The main function of the disconnector is to cut down circuit steadily for safety when servicing high voltage device. However, the corrosion of the disconnector is one of the most easily happening deficiency[2], which will directly cause jam fault of the transmission process and resulted in the operation failure. Meanwhile, the mechanical strength of the parts will lower, and will cause deformation and damage further. Seriously, the safe operation of the power grid would be threatened.

The reason for disconnector corrosion primarily arose form the designing deficiency, the improper material selection, the substandard surface treatment and the unqualified plating procedure, etc[3]. As a result, the material selection problem of the disconnector was discussed in the research. In comparison with the ordinary steel, a special one named weathering resistant steel has been widely used to avoid atmosphere corrosion. The main difference between the weathering resistant steel and the ordinary one is the element content and the corrosion product composition. In special, the weathering resistant steel has a stable and protective inside rust layer whereas the ordinary one does not have. The rust layer can prevent corrosion medium contact with the substrate effectively[4].

As a kind of ordinary steel, the Q235 is widely used in transmission links of the disconnector. Moreover, the Q235NH is one of the weathering resistant steel. The element content of the Q235 and the Q235NH is shown in Table 1.

Table 1. The element content of Q235 and Q235NH.

Steel type	C	Si	Mn	S	P	Cr	Cu	Ni
Q235	0.18	0.25	0.5	0.018	0.016			
Q235NH	≤0.13	0.10~0.30	0.20~0.6	≤0.03	≤0.03	0.4~0.8	0.25~0.55	≤0.65

IOP Conf. Series: Materials Science and Engineering **812** (2020) 012003 doi:10.1088/1757-899X/812/1/012003

For the early stage of corrosion, the composition of the rust layer in both the Q235 and the Q235NH is α-FeOOH, which hardly protect the substrate. Then, some of α-FeOOH transform to γ-FeOOH, which is more stable and has a better protection to the substrate[6]. The essential reason why the Q235NH has a better protection for the substrate is that the element of Ni is in the form of NiFe2O4. It has a high erelectrochemical stability, and resulted in a better protection ability in the rust layer of the Q235NH[8].

In order to evaluate the corrosion resistance of the metal, the corrosion test with artificial acceleration is a good choice. The commonly used methods involves the salt spray test, the gas corrosion test, the damp heat test, the field corrosion test and the cyclic-immersion-corrosion test[9]. Although the field corrosion test is a common way to evaluate the property of weathering resistant steel, it has a very long test cycle. Moreover, the cyclic-immersion-corrosion test can simulate an alternate wetting and drying status, but it also cost too much time[10]. In particular, the salt spray test was used to compare the corrosion resistance of the Q235 and the Q235NH in the outdoor high-voltage disconnector.

1.1 Weight loss analysis

The weight loss analysis is a reliable, effective and widely used quantitative test method to calculate the corrosion rate. It compare the weight of samples before and after they suffer from corrosion[11]. In case of uniform corrosion, the average corrosion rate can be calculated using the weight loss, the area and the test time. Due to the similar density of materials used in the test, the corrosion rate can be calculated according to the follow equation[12]:

$$V = \frac{m_0 - m_1}{At} \tag{1}$$

Where V is the corrosion rate, the unit is g/(m2·h); m0 denotes the weight before corrosion; m1 presents the weight after the remove of the corrosion product; A is the surface area that exposed in corrosive environment of samples; t is the test time.

2.Experiments

2.1 Test samples set

Test samples were metal cylindrical tubes, whose parameters are given in Table 2.

Table 2. Test Samples of Experiments.

Serial number	Material	Outer diameter (mm)	Length (mm)	Wall thickness (mm)	Surface area (mm^2)
1~4	Q235	30	50	3.1	4712.3889
9~12	Q235NH	40	50	2.7	6283.1852

2.2 Sample processing

Samples were polished by 150, 240, 400, 800 mesh sandpaper successively, then immersed in the acetone solution. After that, samples were degreased by the ultrasonic cleaner, followed by a dehydration process with ethyl alcohol and then they were put in drying oven. After 2h, samples were marked 1~4, 9~12 by an oily pen

The analytical balance was demarcated with calibration weights, and then each sample was weighed for three times by the analytical balance. The average of the three values was set to be the initial weight.

The samples' inside space was filled with some material like foam after weighed. Afterwards, the two end of samples was sealed by neutral silicone sealant. These processes made the metal surface exposed in the salt spray environment only with the outside surface. After being sealed, samples were put into the drying oven again for about 8h, and taken out until the sealant was completely dried out.

2.3 Solution configuration

This experiment was performed according to GB/T 10125-2012, four kinds of parameters need to be controlled.

(1) Salt need to be dissolved in the distilled water or the deionized water with the temperature of 25 ±2℃ and the conductivity below 20 μ s/cm. Moreover, the concentration needs to be controlled to be 50g/L±5g/L.

(2) The pH value of the solution that used in the salt spray test requires to be between 6.5 and 7.2, the diluted hydrochloric acid or sodium hydroxide is allowed to adjust the pH value of the solution.

(3) For the collector at any spot of the effective space in this test, the continuous spray collection time is at least 16h, and the average salt spray settlement is 1.2mL at a horizon collecting area of 80cm2 per hour

(4) The temperature of salt spray environment is set to be 35°C.

2.4 Experimental procedure

The experimental procedures are listed below:

(1) Check samples' appearance, take photos.

(2) Before the test, turn on the salt spray test chamber, make it continuous spray for 24h to ensure the stability of the test chamber.

(3) Test the pH value of the salt solution, to guarantee the pH value is within the range of the standard.

(4) The nozzle of the salt spray test chamber is in the middle of the chamber with an umbrella spray, so that when the samples are placed, they are avoided to get too close to the nozzle. Samples are placed at a rigid plastic support, and the salt spray can be freely deposited on the surface of each sample.

(5) When the test comes to 62h, take out sample 1 and 9. Firstly, observe the corrosion morphology and take photos. Secondly, pickle samples in a solution of 500ml hydrochloric acid, 500ml distill water, and 20g hexamethylenetetramine. After the product of corrosion have been cleaned completely, clean the samples with ethyl alcohol. Then put them into the drying oven for 2h and weigh them at last.

(6) When the test comes to 112h, take out sample 2 and 10, do the same operation as step (5).

(7) When the test comes to 162h, take out sample 3 and 11, do the same operation as step (5).

(8) When the test comes to 212h, take out sample 4 and 12, do the same operation as step (5).

(9) Calculate the weight loss data and daw a weight loss curve.

3. Results and discussions

3.1 Corrosion result

The morphology photos of sample 2 and sample 10 are shown in Figure 1, it can be seen from which is as follows:

The Q235 corroded seriously, and was completely wrapped by the rust layer. The surface of the rust layer appeared to be yellow and the inside appeared to be dark brown. After removing the rust layer by descaling liquid, the surface of the metal appeared to be gully-shaped.

Similarly, the Q235NH was wrapped by two rust layers after corrosion, and the surface of the metal after being descaled appeared to be gully-shaped too. Note that the difference between the Q235 and the Q235 NH is that the rust layer in the Q235 NH was more compact.

(a) (b) (c) (d)

Figure1. Morphology photos of samples.
(a) Corrosion morphology of sample2. (b) Corrosion morphology of sample 10.
(c) Derusting morphology of sample 2. (d) Derusting morphology of sample 10.

3.2 Component analysis

Figure 2 shows micrographs of two kinds of material's corrosion surface.

As can be seen from Figure 2, the rust layer of the Q235 after corrosion is powder-shaped and loose. Therefore, it has a poor protective effect on the base material. On the other hand, the Q235NH is flake-shaped, and more compact than its counterpart. Consequently, it has a stronger protective effect on the base material.

3.3 Weight-loss analysis

The weight loss analysis was performed according to GB/T 16545-1996. Table 3 shows the weight loss data of the Q235, the Q235NH and other parameters according to experiment records. The corrosion weight loss curves of the Q235 and the Q235NH are shown in Figure 3.

Table 3. Weight Loss Data and Other Parameters.

NO.	Corrosion time (h)	Initial weight (g)	Final weight (g)	Weight Loss (g)	Weight loss ratio (%)	Surface area (m^2)	Corrosionrate ($g\cdot m^{-2}\cdot h^{-1}$)
1#	62	92.3458	91.9420	0.4038	0.4373	4.7124E-03	1.38
2#	112	93.5709	92.9506	0.6203	0.6629	4.7124E-03	1.18
3#	162	91.4036	90.3074	1.0962	1.1993	4.7124E-03	1.44
4#	212	94.4836	92.8973	1.5863	1.6790	4.7124E-03	1.59
9#	62	116.5849	116.1047	0.4802	0.4119	6.2832E-03	1.23
10#	112	113.1401	112.3063	0.8338	0.7369	6.2832E-03	1.18
11#	162	113.9638	112.4459	1.5179	1.3319	6.2832E-03	1.49
12#	212	117.2886	115.5608	1.7278	1.4731	6.2832E-03	1.30

IOP Conf. Series: Materials Science and Engineering **812** (2020) 012003 doi:10.1088/1757-899X/812/1/012003

(a) (b)

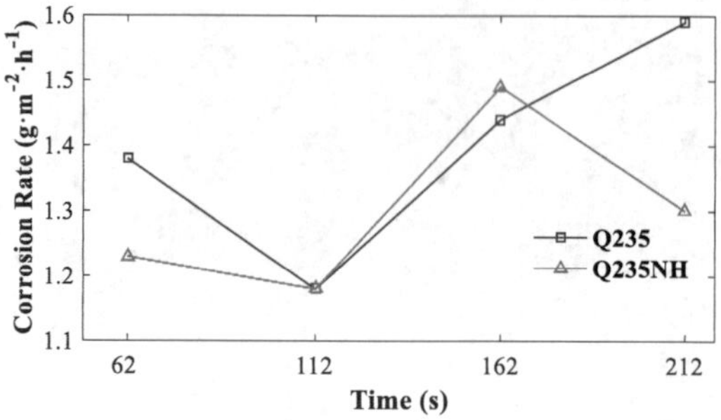

Figure2. Micrographs of Q235 and Q235NH.**Figure3.** Corrosion weight loss curves.
(a) Q235, (b) Q235NH.

Results show that few difference between the Q235 and the Q235NH occured in the early stage of the neutral-salt-spray test. The Q235NH has a certain advantage of corrosion resistance over its counterpart in the later stage. However, the difference is not as good as the data (2~8 times) from related literature. These facts illustrate that the corrosion resistance of the Q235NH is slightly better than that of the Q235 in the neutral-salt-spray test in a short time, and it is necessary to do a longer test to further study the differences between them.

4. Conclusions

a) The rust layer in Q235NH after corrosion is more compact than that of Q235, and leads to a stronger protective effect for the base material.

b) In the early stage of the neutral-salt-spray test, the difference between the Q235NH and the Q235 in is not obvious. However, the Q235NH has a certain advantage of corrosion resistance in the later stage. Accordingly, the Q235NH is a more suitable material for the outdoor high-voltage disconnector.

5. Acknowledgments

This work was supported by the Research Programs of State Grid XinJiang Electric Power Company. (5230DK15001S)

6. Reference

[1] Qiu Z B,Ruan J J, Huang D C,Zhang Y and Zhang E W 2015Mechanical faults analysis of high voltage disconnectors and review of diagnosis technology *High Voltage Apparatus*.**08** pp171-179

[2] Wang H F, Chen G andYu P2009 Improvement measures of isolating switch *Shanxi Electric Power*.**04** pp 28-29+63

[3] Ohyama T,Nagata K and Kanda K 2016 Construction of the seismic-analysis technology of the insulator base metals in a single polestructure disconnecting switch *Electrical Engineering in Japan*. **196** pp 40-49

[4] Xiao K,Dong C F, Li X G and Wang F M 2007 Study on accelerated corrosion tests for carbon steel and weathering steel *Equipment Environmental Engineering*. **03** pp 5-8

[5] Zhang Q C,Wu J S, Zheng W L,Wang J J and Chen J G 2002 The electrochemical characteristics of rust layer formed on theweatheringsteel in cyclic dry/wet laboratory Test *Materials for Mechanical Engineering*. **08** pp 22-24

[6] Wang Z Y, Yu G C and Han W 2004 Investigation On interrelation of indoor accelerated corrosion and atmospheric exposure corrosion of steels *Corrosion Science and Protection Technology*.**16** pp 70-73

[7] Biestek T 1982 Atmospheric corrosion testing of electrodeposited zinc and cadmium coatings. Joha Wiley, New York631

[8] Zhang L, Wang Z Y, Zhao C Y, Cao G W and Liu Y J 2014 State for alloying elements in iron rust using a binary-phase potential-pH diagram and physical analyses *Equipment*

Environmental Engineering. **01** pp 1-6

[9] Qu Q,Yan C W and Cao C N 2003ProgressInexperimental methods foratomspherecorrosion of metals *Corrosion Science and Protection Technology*. **04** pp 216-222

[10] Yin Y Q,Zhao J B,Cheng X Q and Li X G 2012Corrosion rules and mechanism of high strength steel of q450nqr *Science & Technology Review.***16** pp 48-51

[11] Wang Z L, Luo Y,Yang L L, Yin D, Liang H J and Lu H 2013 Comparison of wter-based drilling fluid corrosion evaluation method of loss of weight and chemical *Chemical Engineer,***11** pp 50-52

[12] PXia P,Shan F J, Yang T T,Zheng Y 2015 Corrosion behavior of pure copper in simulated acidRainatmospheric environment *Liaoning Chemical Industry*. **12** pp 1442-1444+1480

IOP Conf. Series: Materials Science and Engineering **812** (2020) 012004 doi:10.1088/1757-899X/812/1/012004

Investigations with Various Inner Shielding Distance Tests for a Novel Coupler-based CPT System Applied for Electric Vehicles Using Electromagnetic Resonant Coupling and Aluminium Shielding Material

J Duan[1,*] and W Wang[2]

[1,2] Department of Engineering and Design, University of Sussex, Brighton BN1 9QT, United Kingdom

*Email: J.Duan@sussex.ac.uk

Abstract. Contactless power transfer (CPT) technology development has been driven rapidly over the past decade by the world-wide trends towards new energy explorations, and numerous reports have been presented in this area. This paper focuses on passive magnetic shielding, which acts as one of the major factors mainly determining the overall CPT system performance when discussing electromagnetic field flux distribution and its real-time effects on magnetic resonant coupling. As a well performance conductive metallic material, aluminium has been adopted to be a passive shielding material in the designed novel H-shape coupler CPT system in this paper, in order to evaluate and find out the optimal inner shielding distance in between the coil and the inner shielding shell. Three inner shielding distances are applied and analyzed across a critical range of system operating frequency, by which the actual CPT system performance differences from perspectives of electromagnetics and power electronics have been illustrated and compared. As a result, it can be noticed that the 15-mm inner shielding gap CPT model is able to yield an optimal system performance with a maximum system efficiency, peak system output RMS power of over 36% and 22 kW, respectively, which also shows an optimal capability to address major concerns over electric vehicle contactless charging. Besides, along with the electromagnetic field parameters generated in the model, such as actual real-time values of flux linkage, magnetic flux density and field strength, it can be found that the 15-mm inner shielding gap prototype is able to achieve better overall magnetic field performance than 5-mm and 25-mm inner shielding distance CPT models.

1. Introduction

Inductive coupling based technologies have been gaining considerable attentions in consumer electronics, factory automations, medical implants, lightings, defense systems and electric vehicles [1]-[4]. Based on inductive coupling principles, contactless power transfer (CPT) has also been proposed more than a decade to mainly address relatively high-power transfer issues and to achieve industrial applications, such as electric vehicles (EV) charging with contactless methods. To cancel out reactive powers from loosely coupler coils to improve system real AC power, compensation topologies have been systematically evaluated in [5]. In this paper, considering inductive coupling via air space or cores could be nonlinear scenarios due to permeability μ and B-H relations of the specific materials used within the coupling module, a proposed H-shape coupler CPT system with aluminium shielding method is introduced to analyze and evaluate the system performance differences in terms of inner shielding distance changes, based on electromagnetics and 3D finite element methods (FEM). Thus, the induced currents, system output RMS real power, and system power transfer efficiency as power electronics indices, along with the flux linkage, magnetic flux density, magnetic field strength and flux line distribution have been investigated in this paper. The overall system performances of 5-mm, 15-mm and 25-mm inner shielding gap CPT models have been analyzed and compared.

2. CPT system design

Based on the effectiveness of high permeability and low core loss of soft ferromagnetic materials, the geometrically improved H-shape soft magnetic core coupler is proposed to form magnetic flux line trajectories and enhance flux distributions within the inductive coupling space in and between the cores and coils, which can lead to optimal CPT system coupling performance especially at electromagnetic resonant coupling status [6]. Figure 1 presents a stationary contactless power transfer system deploying H-shape ferromagnetic cores with a relative permeability μ of 1000, semi-enclosed combined aluminium shielding and four compensation topologies connecting to both sides of the coupling modular. Due to contributions of minimizing reactance magnitude, the S-S compensation topology has been adopted in the designed CPT system in this paper, as shown in figure 2.

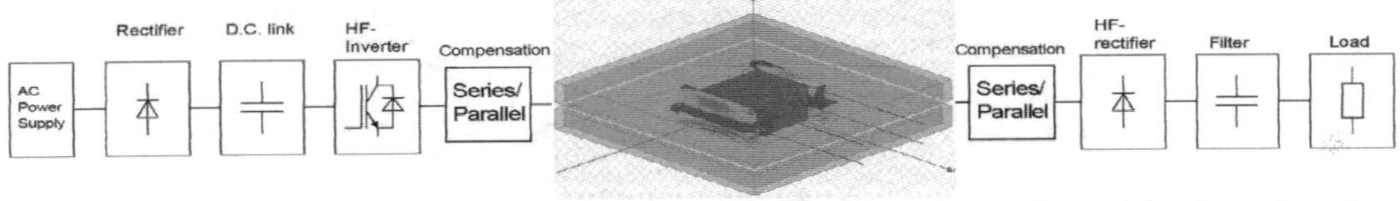

Figure 1. The CPT system configuration with H-shape coupler and combined semi-enclosed shielding.

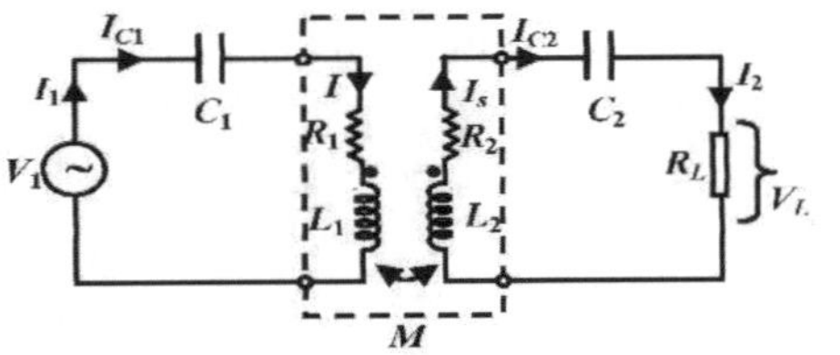

Figure 2. The simplified equivalent circuit of the CPT system with SS compensation topology.

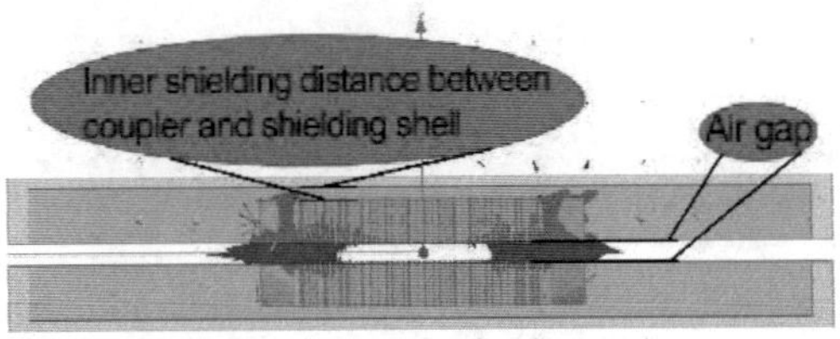

Figure 3. The designed coupler indicating the CPT air gap and the inner shielding distance.

In order to properly produce constantly satisfactory magnetic flux generation with electromagnetic resonant coupling occurring at specific conditions, the innovatively proposed semi-enclosed combined aluminium shielding method has been utilized to further guide and form the actual magnetic flux due to its effects of preventing energy loss in surrounding air [7] and reducing magnetic field flux leakage to the infinite air space [8]. Moreover, different system outcomes of the proposed CPT system with various inner shielding distances, 5 mm, 15 mm and 25 mm, have been delivered and analyzed from both the electromagnetics and power electronics points of view. The designed coupler modular with inner shielding gap and air gap is presented in figure 3. And the detailed design characteristics of the small-sized H-shape coupler CPT system with aluminium passive shielding are given in table 1.

Table 1. The specifications of the designed CPT system.

Parameters	Values
Winding size	100 mm x 100 mm x 20 mm
Core size	150 mm x 150 mm x 20 mm
H-shaped core bar size of each side	150 mm x 15 mm x 20 mm
Primary winding number of turns	80 turns
Secondary winding number of turns	80 turns
Air gaps of the CPT charging system	10 mm
Winding material	Litz wire
Shielding size	400 mm x 400 x 30 mm
Shielding thickness	10 mm
Inner distance between shielding and coil	5 mm
Passive magnetic shielding materials	Aluminium
Relative permeability μ of core material	1000
Relative permeability μ of shielding material	1.000021
S-S compensation capacitors	150 nF
Load resistor	50 ohm
Voltage power supply	RMS 5 kV
Inner shielding distance between coil and shell	5 mm, 15 mm and 25 mm

3. Electromagnetics principles and numerical representations for the designed CPT system

To analyze the CPT system outputs, in electromagnetic field, the electric current density $\mathbf{J}$ induced on the secondary side needs be determined. Hence, the Maxwell equations constitutively expressed by Ampere's law in equation (1), Gauss' law in equation (2), Faraday's law in equation (3) and the B-H curve relation in equation (4) are required to be solved:

$$\nabla \times \mathbf{H} = \mathbf{J} \tag{1}$$

$$\nabla \cdot \mathbf{B} = 0 \tag{2}$$

$$\nabla \times \mathbf{E} = -\frac{\partial \mathbf{B}}{\partial t} \tag{3}$$

$$\mathbf{B} = \mu \mathbf{H} \tag{4}$$

where $\mathbf{H}$ is the magnetic field strength (A/m), $\mathbf{B}$ is the magnetic flux density (T), $\mathbf{E}$ is the electric field strength (V/m), and μ is the permeability non-linearly depending on local value of $\mathbf{B}$ in $\mathbf{B} = f(\mathbf{H})$. The electric current density $\mathbf{J}$ is in A/m^2.

$$\mathbf{B} = \nabla \times \mathbf{A} \tag{5}$$

Simultaneously with Ohm's law in electromagnetics by equation (6) and material equations of Pouillet's law by equation (7), the 3D finite-element method (FEM) is adopted to numerically solve Maxwell equations above and the relative material equations, using the methods of discretizing the modelled 3D space by tetrahedrons, translating the differential equations into algebraic equations [9]. The methods of semi-iterative conjugate gradient for each magnetic vector potential $\mathbf{A}$ [10] and the initial B-H curve corresponding μ can be deployed to determine the flux density $\mathbf{B}$ by equation (5). Repeatedly, the numerical computations can be completed until the convergence of the element permeability.

$$\mathbf{J} = \sigma \mathbf{E} \tag{6}$$

$$\mathbf{R} = \rho \frac{l}{C} \tag{7}$$

where $\boldsymbol{\sigma}$ is the material-dependent parameter conductivity. $\boldsymbol{\rho}$ is the resistivity of the windings. **R** is known as electric resistance of the wire material. $\boldsymbol{l}$ is the total length of the wire. **C** is the cross-sectional area of the wire.

Based on the equations above, equation (1) turns into:

$$\nabla \times (\frac{1}{\mu} \nabla \times \mathbf{A}) = \mathbf{J} \tag{8}$$

$$\mathbf{E} = -\nabla \phi - \frac{\partial A}{\partial t} \tag{9}$$

together with Faraday's law and Lorenz gauge vector potential by equation (9) [11], by using 3D FEM computations, the magnetic flux density **B**, the electric field strength **E** and the magnetic field with values of magnetic flux ϕ in weber can be numerically solved and determined.

In order to determine the self-inductance values of **L** in henry produced by the electromagnetic field on the both sides of the CPT system [12], the relations between the coil current values of i and the flux linkage λ satisfy:

$$\lambda_1 = N_1 \phi_1 = L_1 i_1 \tag{10}$$

$$\lambda_2 = N_2 \phi_2 = L_2 i_2 \tag{11}$$

$$L = \frac{\lambda}{i} \tag{12}$$

To investigate the actual inductive magnetic coupling outcomes and electromagnetic field performance represented and reflected by the numerical vectors and scalars above, the actual RMS power generations given by equation (13) and the efficiencies given by equation (14) and (15) from the front end to the load end of the CPT system are required to be calculated, compared and analyzed. The cos φ is the power factor (PF) caused by phase difference φ between induced voltage and current waveforms.

$$P_{RMS} = V_{RMS}\, I_{RMS} \left| \cos \varphi \right| \tag{13}$$

$$\eta_{Coupler} = \frac{P_{RMS_{Secondary\ coil}}}{P_{RMS_{Primary\ coil}}} \tag{14}$$

$$\eta_{Overal\ system} = \frac{P_{Load}}{P_{RMS_{Power\ supply}}} \tag{15}$$

4. Results and analysis

4.1. Flux linkage and currents with different inner shielding gaps

In this paper, the proposed H-shape coupler CPT systems with aluminium shielding method and three different inner shielding gaps have been investigated across the major significant range of system operating frequency. By emphasizing on the flux linkage in electromagnetics and the induced current in power electronics, it can be seen that the 5-mm, 15-mm and 25-mm inner shielding gap based prototypes produce different characteristic trends against operating frequency.

IOP Conf. Series: Materials Science and Engineering **812** (2020) 012004 doi:10.1088/1757-899X/812/1/012004

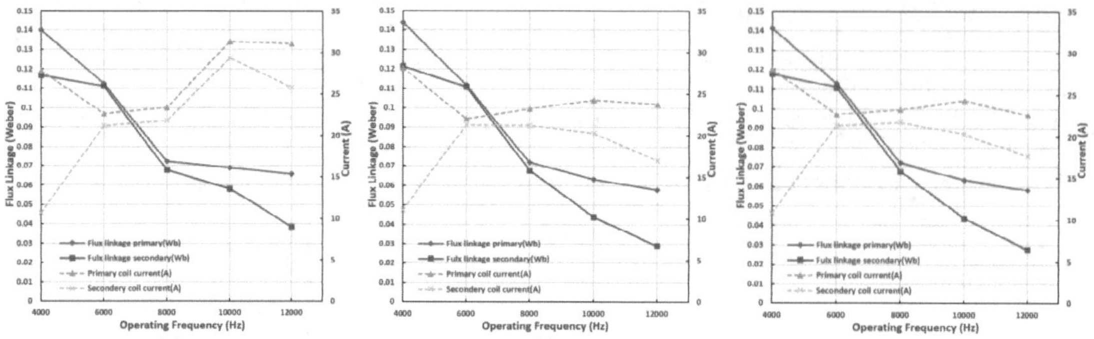

(a) (b) (c)

Figure 4. The flux linkage and current values of the designed CPT system with aluminium shieldingversus system operating frequency. (a) 5-mm inner shielding gap. (b) 15-mm inner shielding gap. (c) 25-mm inner shielding gap.

The results in terms of flux linkage and current have been depicted in figure 4. It can be seen, from figure 4(a), that both the flux linkages of primary and secondary coils tend to drop down against operating frequency despite a more gradual decrease after the frequency point of 8 kHz. In the meantime, the primary coil current shows a slight decrease before 6 kHz and then tends to rise up to a peak value of 31.284 A at 10 kHz. The secondary coil current shows an increasing tendency before 10 kHz as well, reaching its maximum value of 29.292 A. However, the secondary coil current appears to drop down after 10 kHz point, which could be corresponding to the maximum system performance point along with the calculated resonant frequency for the 5-mm inner shielding gap model.

By comparing figure 4(b) and (c), it can be noticed that both the 15-mm and 25-mm inner shielding gap models show very similar tendencies upon flux linkages and coil currents, which may reflect that 15 mm for inner shielding distance could be a critical point due to no further significant differences regarding flux linkage and current against operating frequency with other parameters kept the same in the designed CPT system. Nevertheless, the primary coil currents of both 15-mm and 25-mm inner shielding gap models appear to show maximum values of 24.278 A and 24.296 A at 10 kHz. While, the secondary coil currents of both the 15-mm and 25-mm inner shielding gap models tend to achieve optimal values between 6 kHz to 8 kHz, which can reflect that inductive couplings with satisfactory induced currents to secondary side of the system could be reached during this frequency range in order to produce higher real power on the load end.

4.2. Efficiencies and RMS real powers with different inner shielding gaps

The actual real-time phase angles of induced currents and voltages are changeable in both sides of the system, which determines different real-time power factors (PF) and generates variable RMS real power results [6]. After each CPT system model tends to produce AC stabilized output waveforms, the results of RMS power and efficiency can be computed based on equation (13)-(15) in order to be analyzed.

As indispensable indices to evaluate and reflect the CPT system performance, as shown in figure 5, three different inner shielding gaps based models yield different RMS powers and efficiencies. From figure 5(a), it can be seen that the overall system efficiency shows an increasing tendency to reach a maximum value of 27.480% at 10 kHz. At the same operating frequency, the system output power achieves its peak value of 42.899 kW, which indicates the magnetic resonant coupling status tends to occur at its approximate calculated natural resonant frequency of 10 kHz.

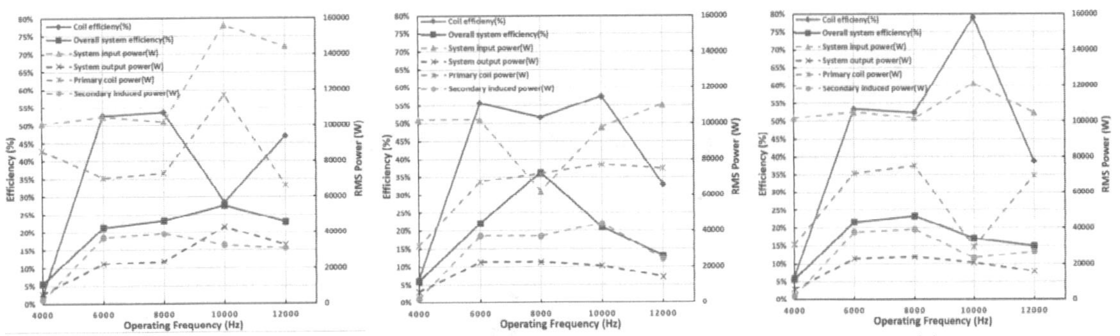

(a) (b) (c)

Figure 5. The RMS real powers and efficiencies of the designed CPT system with aluminium shielding versus system operating frequency. (a) 5-mm inner shielding gap. (b) 15-mm inner shielding gap. (c) 25-mm inner shielding gap.

Regarding the 15-mm inner shielding gap CPT model, it can be noticed, from figure 5(b), that the overall system efficiency shows a sharp peak point at operating frequency of 8 kHz, which indicates a maximum value of 36.257%. Nevertheless, the peak value of the system output power is produced at 6 kHz, which is 22.626 kW. Significantly, it can be found that, across a wide range of operating frequency after 6 kHz, the system output power is able to stay highly satisfactory and sustainable with values of over 20 kW, which reflects that the acceptable electromagnetic field flux distributions and resonant coupling performance in this 5-mm inner shielding gap model enable stable and satisfactory power transfer ratings and overall system efficiencies for the CPT applications to electric vehicles.

In terms of the outcomes of the 25-mm inner shielding gap CPT prototype depicted in figure 5(c), the overall system efficiency tends to be achieved at 8 kHz with a resulted value of 23.245%. In the meantime, the system output power reaches its peak value of 23.676 kW at 8 kHz. By comparison, it can be also found that the average values of overall system efficiency against each operating frequency are almost slightly lower than those of the 5-mm inner shielding gap CPT prototype. Significantly the maximum system efficiency of the third model is much lower than the performance of the 15-mm inner shielding gap CPT prototype, by about 13%, which can reflect and prove that the 15-mm inner shielding gap between coil and inner shell surface can be the optimization for an aluminium passive shielding method.

4.3. Analysis and comparison on main results and field performance with different inner shielding gaps

In addition to graphic tendency comparison and analysis, the detailed major numerical results computed based on theories and equations in previous sections have been illustrated in table 2 for comprehensive evaluations. As can be seen, the overall system efficiency of 15-mm inner shielding gap model shows the best outcomes within these three different inner shielding gap models although the maximum system output power of the 15-mm inner shielding gap model is about 20.27 kW lower than 5-mm inner shielding gap model, which could illustrate and reflect that the latter model can deliver faster charging due to its higher output real power on load. Nevertheless, the 15-mm inner shielding gap model seems more able to produce better efficiency of energy utilization with 36.26% as a peak point. On the other hand, 25-mm inner shielding gap model tends to generate slightly weaker performance than 15-mm one, which could also mean that 15 mm can be the optimal inner shielding gap for the designed CPT system from system efficiency point of view.

Table 2. Main numerical results of the CPT system with three inner shielding distances.

Operating Frequency (Hz)	System with 5-mm inner shielding gap			System with 15-mm inner shielding gap			System with 25-mm inner shielding gap		
	System Output Power (W)	Coil Efficiency	Overall System Efficiency	System Output Power(W)	Coil Efficiency	Overall System Efficiency	System Output Power (W)	Coil Efficiency	Overall System Efficiency
4000	5659.8033	2.2591%	5.6023%	6093.5015	6.6358%	5.9602%	5802.1568	6.4614%	5.7001%
6000	22389.6902	52.7415%	21.3399%	22626.4968	55.6789%	22.1569%	22734.2756	53.4214%	21.6117%
8000	23704.4088	53.6170%	23.2055%	22526.4449	51.8073%	36.2570%	23676.0812	52.3811%	23.2451%
10000	42899.5869	28.3367%	27.4804%	20566.4916	57.5064%	20.9863%	20612.5843	79.1434%	17.0018%
12000	33008.5921	47.0085%	22.9074%	14401.9097	32.9440%	13.0551%	15617.5647	38.6898%	14.9076%

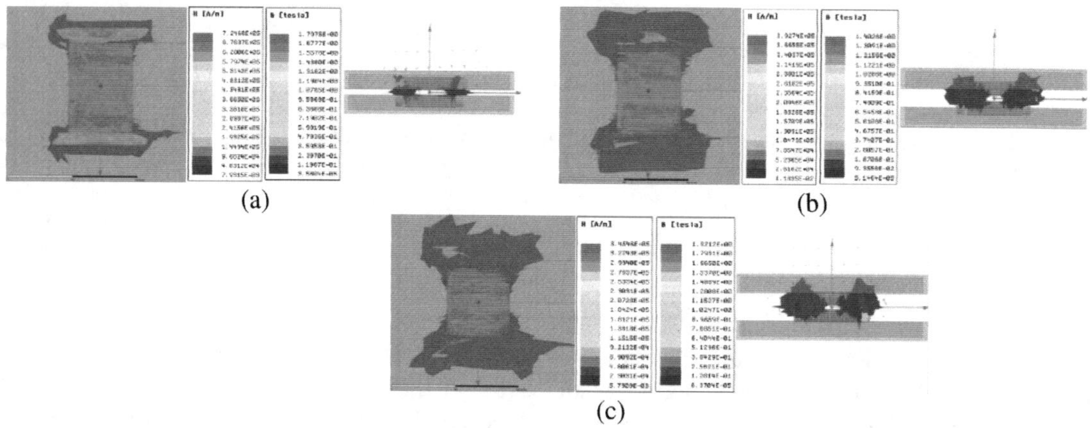

Figure 6. The electromagnetic field distributions of the designed H-shape coupler CPT system with aluminum passive shielding at 1 ms, indicating the magnetic field strength **H** *(A/m)* and the magnetic flux density **B** *(Tesla),* at approximate resonant coupling status as an optimization.(a) The top and left view of the field for 5-mm inner shielding gap. (b) The top and left view for 15-mm inner shielding gap. (c) The top and left view for 15-mm inner shielding gap.

From the perspective of electromagnetics, the magnetic field strength **H** in A/m and the magnetic flux density **B** in Tesla can reflect the field performance of a system based on Maxwell's equations. As can be seen from figure 6, different field flux distributions and parameters are shown to be analyzed. The 5-mm inner shielding gap system can produce maximum **H** and **B** values of 7.2468 x $\mathbf{10^5}$ A/m and 1.7975 Teslas, respectively, at some crucial surface edges and corners of the coupling modular when the system efficiency reaches its maximum of 27.48%. The 15-mm inner shielding gap system can generate maximum **H** and **B** values of 3.9274 x $\mathbf{10^5}$ A/m and 1.4026 Teslas, respectively, when the system tends to be magnetically resonant coupled at 8 kHz as presented in figure 6(b) and table 2. From figure 6(c), it can be seen that the peak values of **H** and **B** field can be 3.4546 x $\mathbf{10^5}$ A/m and 1.4026 Teslas, respectively. Comprehensively, the overall vector values throughout the entire coupling space of the 15-mm inner shielding gap CPT system can be found to be optimized although its **H** and **B** maximum points are slightly weaker than those of the 25-mm inner gap model. Therefore, the 15-mm inner shielding gap can be an optimization from the perspective of electromagnetics.

5. Conclusion

This paper presented a small-sized geometrically improved H-shape coupler CPT system and analyzed performances of using three different inner shielding distances. By focusing on RMS power transfers, CPT system efficiencies, generated electromagnetic field characteristics, it can be found that the 15-mm inner shielding gap can be an optimal shielding method with maximum outputs when considering system power transfer, system efficiency and electromagnetic coupling performance. The 15-mm inner shielding gap CPT system can produce a maximum system efficiency of 36.26%, a peak system output power of 22.63 kW, satisfactory values of magnetic field strength and flux density throughout the coupling modular at about the magnetic resonant frequency.

6. References

[1] Pengfei Li, R. & Bashirullah, 2007. A Wireless Power Interface for Rechargeable Battery Operated Medical Implants. IEEE Transactions on Circuits and Systems II: Express Briefs, 54(10), pp.912–916.

[2] Kawamura, A., 1996. Wireless transmission of power and information through one high-frequency resonant AC link inverter for robot manipulator applications. IEEE Transactions on Industry Applications, 32(3), pp.503–508.

[3] Robertson, D. et al., 2011. High power IPT stage lighting controller. 2011 IEEE International Symposium on Industrial Electronics, pp.1974–1979.

[4] Hui, S.Y.R. & Ho, W.W.C., 2005. A new generation of universal contactless Battery Charging platform for portable Consumer Electronic equipment. IEEE Transactions on Power Electronics, 20(3), pp.620–627.

[5] Sohn, Y. et al., 2015. General Unified Analyses of Two-Capacitor Inductive Power Transfer Systems: Equivalence of Current-Source SS and SP Compensations. IEEE Transactions on Power Electronics, 30(11), pp.6030–6045.

[6] Duan, J., Wang, W., 2018. Electromagnetic Coupling Optimization by Coil Design Improvements for Contactless Power Transfer of Electric Vehicles. In Proceedings of the Future Technologies Conference (FTC) 2018. Advances in Intelligent Systems and Computing, vol 881, (2018) pp. 944-958. Springer, Cham, Switzerland. DOI= https://doi.org/10.1007/978-3-030-02683-7_69

[7] Paul, C.R., Introduction to electromagnetic compatibility, New York: Wiley, 1992.

[8] Wohlfarth, E.P., Ferromagnetic materials: a handbook on the properties of magnetically ordered substances, Amsterdam: North-Holland, 1982.

[9] Sadiku, M., Numerical techniques in electromagnetics, Boca Raton: CRC P, 1992.

[10] Silvester, P., Ferrari, R., Finite elements for electrical engineers (2nd ed.). C.U.P, 1990.

[11] Jackson, J.D. & Okun, L.B., 2000. Historical roots of gauge invariance. Reviews of ModernPhysics, 73(3), pp.663–680.

[12] Grover, F., Inductance Calculations: Working Formulas and Tables. Dover, 1962.

IOP Conf. Series: Materials Science and Engineering **812** (2020) 012005 doi:10.1088/1757-899X/812/1/012005

The Research on Data Centre Construction Invested by Power Grid Company Mode

Cheng Wang[1], Hongjian Zhu[2], Zuquan Liang[1]

[1] State Grid Sichuan Comprehensive Energy Service Co.Ltd, Donfeng road No 21in Jinjiang district of Sichuan Province Chengdu City, Chengdu City, 610016, China
[2] State Grid Panzhihua Power Supply Company Panzhihua corporation, Middle part of Panzhihua Road in Sichuan Province Panzhihua City, Panzhihua City, 617000, China

Email: Liangzuquan@126.com

Abstract. Data Centre is one of the hardware facilities of the present stage social development trend. It will be rapid developed in the next few years. While not only its construction cost is high, but its operation cost is high. The State Grid Company is undergoing a transformation and become to a comprehensive energy service provider instead of power suppliers. It has the advantages in the aspects in cost, power supply reliability and operational maintenances if it takes part in Data Centre construction. The investment is in line with the development strategy of SGC, belongs to the high quality of investment projects and it is the new investment direction of SGC.

1. Introduction

With the development of cloud computing, big data, the source of data for the Data Centre has been extensive more and more. The data equipments are accessed diversified more and more and the data capacity has become larger whose process need to be rapid and efficient more and more. All of these made it definitely puts forward higher requirements to the Data Centres. In China, the Data Centre is mainly on the Internet and communications enterprises. The State Grid Company (SGC) has established some platform Data Centres only in the need of business such as own SG186 system, ERP system. The SGC is promoting the fusion of the Internet and the State Grid, building the Energy Internet and ensure itself to be a world-class Energy Internet company with global competitive abilities. It has announced that the SGC must set up the "Ubiquitous Power Internet of Things" within 3 years. So it is inevitable to strengthen the construction of its own Data Centre not only the SGC but also every provincial Power Grid Companies.

2. The data centre

2.1. The characteristics of the data Centre

The characteristics of the Data Centre which are the most important requirements are:

1. The energy must be enough.

Like many Data Centre constructions in our country are built in the northwest and southwest where are full of the coal, water, wind and light energies corresponding.

2. The Data Centre should be stable and reliable.

To the Data Centre, the environment temperature should be stable and the requirements of power supply quality should be very high. The safety, stability and reliability of power supply are the top priority[1].

3. The Data Centre should be energy conservation and environmental protection.

The operation cost of Data Centre is very high. About 50%-70% of it is electricity which includes the cost of equipment utilization, fan, air conditioning, etc.

It is shown that the construction safety, easy maintenance of power supply system can enhance its competitiveness and the development trend of Data Centre must be green Data Centre.

2.2. The market scale is huge

There are nearly 1/3 large Data Centre in China focused on the Beijing, Shanghai and Guangzhou city. It is the same as the global trend that the Data Centres are mostly centralized in the economic developed area. The market participants include the telecom operators (China Telecom, China Mobile and China Unicom) which are the main force of the construction of the Data Centre. Each of them has high significant advantages in aspect of policy, capital; human resources and more than 60% of the Data Centres are held by the three major telecom operators.

According to the China IDC data, the market scale of the IDC will be close to 190 billion Yuan in 2019. In 2016-2019, 4 year compound growth rate will be 21.1% which is significantly ahead of the international level. The core driving force to the market scale growth is the demand of function outsourcing stable of enterprise IT and Vertical industry customers such as the Cloud services, mobile Internet, video, online games.

Vertical industry customers such as, and function of enterprise IT outsourcing stable growth of demand, is the core of pulling the size of the market driving force [2].

From 2015 to 2020, the compound growth rate of cloud services market in China is 29.4% while the overall market scale accounts for only 4% of global cloud services market and it has larger development potentiality. As the Alibaba Cloud is the leading enterprises of domestic cloud services in China, the 10-year-compound growth rate of it will be 56.5% from 2015 to 2025 which is forecasted by Credit Suisse AG.

The high improvement of new market scale is also benefit from the increasing coverage and declining rates of 4G networks, which makes the 4G network user coverage growth rapidly. This is also greatly promoted the video consumption and the increase in data traffic. In the future, as the significance revolutionary of 5G mobile communication network researches and application which is lead by China, the mobile Internet customer demand is expected to have a more substantial growth.

In addition to the demand of mobile Internet, more and more enterprises pay attention to the use of the Data Centre. Each IT function outsourcing of big enterprise can make the enterprise focus much on the development of its core business and at the same time reduce its costs. Compared to traditional industry, the financial enterprises care about data security and recovery requirements higher and it make them prefer to the high performance and reliable Data Centre.

2.3. The growth of market supply sides goes slowly

While the demands are continues to increase, the growths of market supply sides have fallen sharply and the products demanded exceed supplied. The Data Centre supply is only 12.4% in 2016 and it is expected to further reduce. Although the growth is high, the stock scale is still small, less than a quarter of the United States.

In summary, there are 3 aspects that why the growth of supply-side is slowdown:

1. The energy consumption controlled by government.

The power consumption of Data Centre is much higher than ordinary buildings as the power supply system of Data Centre should support a large number of servers and their attendant equipment together.

To the regions especially the major cities in China which are short of electric power, both the national development and reform commission and the State Grid have greatly limited the number and scale of Data Centres for caring about the examination and approval of energy consumption.

2. Secondly, there is another important factor that the urban land of first-tier cities is limited. As now the Data Centres are mainly small and medium-sized, the unit operation cost is high and can't achieve scale effects. The urban land is now supply structural imbalanced while the first-tier cities and city core network node is especially significant.

3. The comprehensive energy service mode

3.1. The comprehensive energy service

The comprehensive energy service is a new type of service to satisfy the diversified energy production and consumption of clients. It includes such as energy Resources planning and design, project investment and construction, diversified energy services and investment and financing service operation etc.

The comprehensive energy service is an advantage service mode. It is conformed to the rapid development of our country's economy and the improvement of residents' living standards. Now more and more power users are paying attention to the quality of electric power. On supply side it is necessary that the structural reform of electric power and electric power service requirement has quietly shifted from "quantity" to "quality".

3.2. The kinds of services

At present, to carry out comprehensive energy services has become an important development direction of energy power companies.

It has several basic kinds of services as follow:

1. Comprehensive supply of power

There are some professional terms of the comprehensive supply of power such as Products integration, Channel integration and Distributed multi-energy complementary system. The Products integration means provide terminal with coal, electric power, coal gas and other energy products at the same time; The Channel integration means operating a variety of energy network at the same time. The different energy sources can be integrated into the goal district and supply energy with single channel, so the Distributed multi-energy complementary system means solving demand of the gas, refrigeration, heating, electricity and other energies together[3].

2. Improve the energy efficiency, reduce the energy cost

It means to promote of the energy-saving equipment, the implementation of energy saving transformation on “Hardware” and optimize the energy solutions, improve the consumer combination, etc on “software”.

3. Equipment maintenance

The market entry point is the traditional customer services such as equipment maintenance and so on which is relied on the advantages of technologies and equipments. Diversified business of Customer needs can realized through the methods like monitoring, energy saving transformation, alternative energy, energy custody etc.

4. Cleaned Micro-Grid

The Micro Grid which is integrated with Consumption and Energy Supply is supported by distributed new energy, energy storage technologies[4] which is indicated by Wu Geng, Wu Qingguo, and Wang Haojing(2019).

5. Intelligent Home

Based on "Intelligence+ electrification", power supply scheme is proposed for full living service design and retrofit which is covering equipment fabrication, intelligence control system as such.

Many large foreign energy power enterprises focus on the 3, 4 and 5.

4. The significance and advantages

4.1. The strategic goal of company

The SG Company has been clear about the strategic goal is to: build globally competitive world-class energy Internet companies. The Data Centre is one of the bases of Internet Companies. In recent, the No. 1 files of State Grid [2019], clearly put forward: "In 3 years of efforts, strive for and preliminary built to the globally competitive world-class energy Internet companies before 2021"; "Research on the new model which the explore the use of substation resources construction operation filling in electricity (energy storage station) and Data Centre, ... Promote collaborative development and mutualism between the emerging business and grid business".

4.2. The advantages of cost

Electricity is the main operating cost of the Data Centre, and the grid company can reduce and control the expenditure through buying/selling electric mode. So the cost advantage is obviously than other enterprises on electricity which is an important and basic economic indicator.

4.3. The advantages of technologies

The SG Company is a professional company of electric power with the incomparable advantages such as technologies, capital, talents etc. It can supply the best electric power and has advantage in aspects of equipment operation and maintenance. It also can settle the electric power quality problem immediately. The SG Company has the powerful abilities to settle the electric power quality problem immediately.

4.4. The advantages of land

According to the scale of substation at present and the equipments are integrated more and more, it can make full use of existing land of substation to construct the Data Centre, avoid the influence factors such as land acquisition and can layout rapidly.

4.5. The advantages of cooperation mode

Data Centre belongs to heavy Fixed Assets Investment. The service object is different with different construction purpose. As its high construction and operational cost, many domestic enterprises may not consider to build it while take into account to implement the requirements by the way to rent.

5. Investment and Profit

Refer to the online information; a Data Centre of some industrial park was built as the large Data Centre construction and the overall planning construction area is 30000 ㎡. It was designed to hold about 3000 cabinets and total investment is 700 million Yuan. According to the overall planning and implemented by Stages, the first construction scale is 300 standard cabinets which will increase with the future market demand gradually, eventually it reach to 3000.

In No.1 stage, the investment of Data Centre is 65 million Yuan. In first year, 300 sets of equipments and system engineering cabinets including power supply, refrigeration, fire control, security, etc. should be equipped. 1200 sets of servers should be configured. It needs to invest 222.95 million Yuan and the cloud platform constructions cost 25 million Yuan. The total cost is 312.95 million Yuan. The basic operating cost is 58.96 million Yuan a year.

General the economic benefits of the Data Centre can be realized by two rental ways which are renting the cloud platform and rental cabinet. As the12000 cloud servers are used some cabinets, there are about 1500 sets of cabinets can be provided to rent. As the price of each rack leasing market is about 120000Yuan per year, the annual revenue could be 180 million Yuan.

6. Conclusion

The Data centre will be rapid developed in the next few years and is one of the hardware facilities of the social development trend at present. It is proved that the investment of Data Center construction is in line with the development strategy of SGC and it belongs to the high quality of investment projects. The investment scale is larger and the income is stable, and companies involved in certain advantages, cost and construction suggest that it is one of the best projects to implement when SGC is transformed from a comprehensive energy service provider instead of power suppliers.

7. Acknowledgments

The article is supported by the State Grid Sichuan Comprehensive Energy Service Co.Ltd,. We would also like to thank some colleagues of the State Grid Sichuan Comprehensive Energy Service Co.Ltd, for their valuable feedback.

8. References

[1] Shi guowei (2019) “The analysis of cloud computing Data Center construction pattern”, China Management Informationization 22(10) (2019): 86-87.

[2] Yu Jiangtao(2019). "Research on Planning and Construction of Data Center Computer Room",China Computer & Communication 10 (2019): 231-232.

[3] Lin Li, Cai Xuexuan(2019). "Source-load Cooperative Operation Strategy of Integrated Energy Service Provider Based on Wind Power Accommodation Demand", Power System Technology 43(7) (2019): 2517-2527.

[4] Wu Geng, Wu Qingguo, and Wang Haojing. etc. (2019) "Energy purchasing strategy of multi-energy service provider considering risk aversion", Electric Power Automation Equipment 39(6) (2019): 12-20.

IOP Conf. Series: Materials Science and Engineering **812** (2020) 012006 doi:10.1088/1757-899X/812/1/012006

The technology of using inflatable structures made of special materials to adapt the de-orbiting system for removing spacecraft

V V Koryanov[1,a], A G Toporkov[1,b], V P Kazakovtsev[1,c], A A Nedogarok[1,d]

[1] Bauman Moscow State Technical University, 5, b1, 2-nd Baumanskaya Street, Moscow, Russian Federation, 105005

Email: [a]vkoryanov@mail.ru, [b]toporkov.90@mail.ru, [c]kafsm3@bmstu.ru, [d]nedogarok.aa@bmstu.ru

Abstract. We are today facing a huge issue on the concentration of space debris in near-Earth space. More than 15,000 debris have been recorded, and this represents a danger for current and future missions to space. A solution to this issue is to remove these spacecrafts from orbit. One of the ways to do it is to use inflatable device. These braking device can be installed on the landing vehicles on Earth, before their launch, or after they have been labelled as space debris. Studies have already been conducted about inflatable device. The results of these studies show that using an inflatable device on an orbital spacecraft significantly reduces its lifetime (it then does not exceed 25 years in most cases). Using inflatable device instead of rigid device has advantages, such as a small mass and a great compacity, which is convenient to respect the required launching volume.

1. Introduction

Currently, the number of devices launched into space is constantly increasing. Spacecraft may fail for one reason or another, or they may run out of life. They remain to fly in orbit, becoming space debris. Various measures, both administrative and technical, are being taken to combat space debris. One of the options that can be offered is to use inflatable braking devices for converting spent spacecraft, which must be laid into the design of spacecraft. At the last stage, when the spacecraft becomes space debris, an inflatable braking device unfolds, which, by changing the geometry of the body, gives impetus to the spacecraft and the spacecraft begins its movement from orbit along a ballistic trajectory. Here you can consider two options for the development of further events. The first option is movement in a ballistic orbit, when the spacecraft burns in the atmosphere, the second case is movement along a ballistic trajectory, in which the spacecraft performs a successful landing on the surface using inflatable braking devices.

The authors of the paper consider the most to consideration both options. The paper is devoted to the creation of such a structure, the use of materials for such a design of inflatable braking devices, which would allow to open up in space and fulfill the assigned mission. The main types of forms that can be applied for such a purpose are considered.

Previously, landing issues were considered by the authors in [1, 2, 3].

The prototype for such a proposal in the use of inflatable braking devices is the MetNet project [4] and one of its continuations - the RITD project [5].

2. Review of materials for such a shell

Shell. Mass industry offers a huge amount of film materials with different composition, thickness, purpose. Film materials were selected according to the following criteria:

1. Preservation of mechanical, chemical and electromagnetic properties in the temperature range of the stay of a spacecraft on the Earth's orbit (-150 ... 250 °) during the entire lifetime of the spacecraft and its subsequent de-orbit.
2. Flexibility and strength sufficient to preserve the integrity and tightness of the shell after installation, pressing into the compartment, storage, subsequent disclosure and pressurization.
3. Low porosity.
4. Manufacturability cutting and fasteners. For example, weldability for the possibility of soldering hermetic joints, or ease of cutting.
5. Mass production and reasonable cost.

Polyethylene terephthalate. The material is also known as polyethylene glycol terephthalate, PET, PETG, polyester glycol, mylar. Non-toxic. Well cooked. Mass produced and finds extensive use in the national economy from films, sheets, hoses, to bottles, cases and other products. Relatively hard to break (destroyed at 2-4% deformation).

PET is often the basis for a multi-layer metallized film used to limit radiant heat transfer on spacecraft. Film porosity varies depending on thickness, supplier and other factors.

The general purpose film produces a large number of domestic and foreign manufacturers.

Polyimide (Kapton). Among the advantages of this polymer, in addition to chemical and temperature resistance, high radiation resistance is distinguished, however, it is unstable to alkalis. In particular, NASA viewed it as the preferred material for solar sails [6]. Depending on the variety, the working temperature range can reach -260 ° C to + 220 ° C. The review revealed an industrial release of film with a thickness ranging from 30 microns. Kapton is used in special industries (electronics, energy, aerospace), as well as in construction.

Teflon (PTFE). Organofluorine compound. Heat-resistant, welded, not flammable. Chemically neutral, films have low porosity. Finds a huge number of applications in industry. In the Russian Federation, fluoroplast properties are standardized according to GOST 24222-80. More plastic to break than PET (breaks at 50 - 200% strain, depending on the grade). Accordingly, films with a thickness ranging from 0.1 mm were found on sale.

Graphene film. Graphene is a modification of solid carbon in which atoms are lined up in layers, with each layer representing a repeating structure in the form of carbon atoms lined up in the form of regular hexagons.

Currently an experimental material. It is not produced in large quantities and is used mainly in the chemical and electronic industry. However, potentially producing films in large formats for mechanical applications. This has the record withstand tensile stresses.

GrapheneX Group (https://www.facebook.com/GrapheneXDYT/) from the University of Delft (Delft University of Technology), the Netherlands, since 2015 has been developing and testing elements of spacecraft, the design of light sails based on the use of graphene.

3. De-orbiting device configurations

De-orbiting devices can be designed in shapes of sphere, thorus, cone, pyramid, either dome, plane etc. The development of inflatable structures is covered in other articles [7, 8, 9].

According to the newthonian model for low-density high-velocity flow gas particles impact surface of de-orbiting device.

Body get the momentum component normal to it's surface.

The result force applied to area S is equal

$$\vec{R}_a = -\vec{n}V_\infty^2\rho \cdot S\sin^2\alpha = -\vec{n}\left(\vec{V}_\infty \cdot \vec{n}\right)^2 \rho \cdot S, \quad (1)$$

Where:

$\vec{n}$ – unit vector normal to surface S.

V_∞ – gas partickles velocity, equal to spacecraft velocity relative to moving atmosphere V_{atm}, with minus sign,

ρ – atmosphere density,

α – angle between flow velocity vector and surface.

Drag force $\vec{X}_a$ respectively is equal

$$\vec{X}_a = -\frac{\vec{V}_\infty}{V_\infty^2}\left(\vec{n}\cdot\vec{V}_\infty\right)^3 \rho \cdot S, \quad (2)$$

or

$$X_a = \frac{1}{V_\infty}\left(\vec{V}_\infty\vec{n}\right)^3 \rho \cdot S, \quad (3)$$

Spacecraft velocity relative to atmosphere, rotating with Earth (airspeed)

$$\vec{V}_{\text{atm}} = -\vec{V}_\infty = \vec{V} + [\vec{r}\times\vec{\omega}_3], \quad (4)$$

$\vec{V}$ – spacecraft velocity in ECSF reference frame,

$\vec{r}$ – spacecraft position vector in ECSF reference frame,

$\vec{\omega}_3$ – Earth rotation angular velocity vector.

Drag force is usually expressed through dynamic pressure q_∞ and drag coefficient c_{xa}, or through ballistic coefficient σ, respectively

$$X_a = c_{xa}\cdot q_\infty \cdot S, \quad (5)$$

where $q_\infty = \frac{\rho V_\infty^2}{2}$, or

$$X_a = m\cdot\sigma\cdot V_\infty^2\cdot\rho. \quad (6)$$

From (5) and (6)

$$\sigma = \frac{c_{xa}S}{2m}, \quad (7)$$

m – spacecraft mass.

From (3) and (5) c_{xa} for plane surface

$$c_{xa} = \frac{2}{V_\infty^3}\left(\vec{V}_\infty\vec{n}\right)^3 = 2\cos^3\left(\widehat{\vec{V}\vec{n}}\right), \quad (8)$$

and ballistic coefficient

$$\sigma = \frac{1}{m}\cos^3\left(\widehat{\vec{V}\vec{n}}\right)\cdot S. \quad (9)$$

Calculational software was developed in appliance with newthonian flow theory, which used for estimation of drag and ballistic coefficients for different configurations of deorbiting device and their orientation in the flow. The data are listed below.

For example, *the cone design* is compact in the folded form and provides a large area in the expanded form. The conical shape facilitates the angular stabilization of the spacecraft near the zero angle of attack.

An example of the polar diagram of the dependence of c_{xa} on the direction of blowing for a cone with an half angle of 30° is shown in Fig. 1.

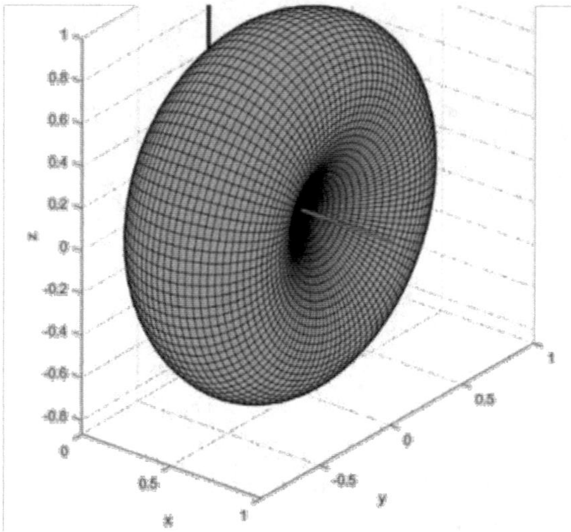

Figure 1. Polar diagram of the dependence of the drag coefficient of the cone on the direction of the speed of the oncoming flow

The maximum drag coefficient in the front hemisphere c_{xa}= 1.05 is achieved with the direction of motion perpendicular to the cone generator, the minimum c_{xa}= 0.505, when moving with the toe in the direction of the velocity vector.

The tetrahedral shape of the braking device can be supported by deployable reinforcing elements, for example, wire or strip ribs unwound from the coil. This design does not need a pressure generator and cannot be subject to leakage if the film sheath is damaged, but has less stability and requires the presence of drives for mechanical deployment. The tetrahedral shape contributes to the angular stabilization of the spacecraft near the zero angle of attack.

An example of the polar diagram of the dependence of c_{xa} on the direction of blowing for a cone with an half angle of 30° is shown in Fig. 2.

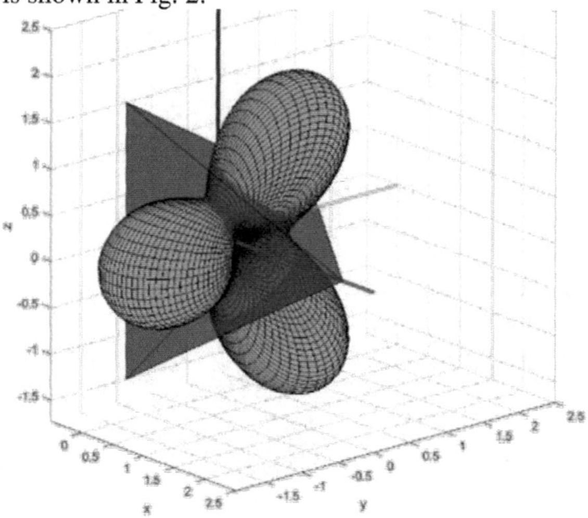

Figure 2. Polar diagram of the dependence of the coefficient of drag of a tetrahedron on the direction of the speed of the oncoming flow

The maximum drag coefficient in the front hemisphere c_{xa} = 2 is achieved with the direction of the incident stream vector perpendicular to the lateral face of the tetrahedron, the minimum c_{xa} = 0.213 - when moving with the toe in the direction of the velocity vector.

The brake device is *a flat frame* of various shapes, on which the film is stretched. Has the greatest drag coefficient c_{xa} = 2 (see Fig. 3).

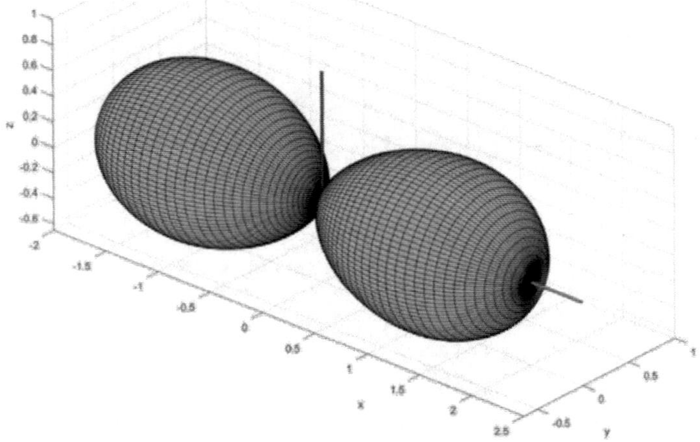

Figure 3. Polar diagram of the dependence of the coefficient of the drag of the plane on the direction of the speed of the oncoming flow

The maximum drag coefficient in the front hemisphere c_{xa} = 2 is achieved with the direction of the vector of the incident flow perpendicular to the plane of the brake.

The cupola device is a parachute. The film canvas is stretched by means of unfold able wire "slings" to maintain the maximum area and radius of curvature. Unlike a flat device, does not have a reinforcing frame around the perimeter, so under the influence of the oncoming flow the shape can differ substantially from the flat one and have a radius of curvature. However, in comparison with a flat device, the structure and the deployment scheme are simplified, and the pressure of the oncoming flow naturally contributes to maintaining the shape of the cupola and stabilizing the objects.

Spherical braking device is structurally the simplest and does not require reinforcing elements. It can be a thin-film ball, supercharged by a gas generator. Has the same drag coefficient c_{xa} = 1 for any orientation of the objects.

There are three main variants of the mutual placing of the braking device and the satellite is shown in Fig. 4. If there are no projecting parts from the side of the braking device, capable of damaging it, the most effective placement is assumed to be closely. This arrangement prevents the satellite from spinning with respect to the braking device and possible impacts on it.

If there are projecting parts on the body of the spacecraft from the opening side, the braking device can be carried out on a remote on a hard construction. However, with such a scheme, the braking device is less protected from damage when hits the housing, the overall rigidity of the structure is reduced.

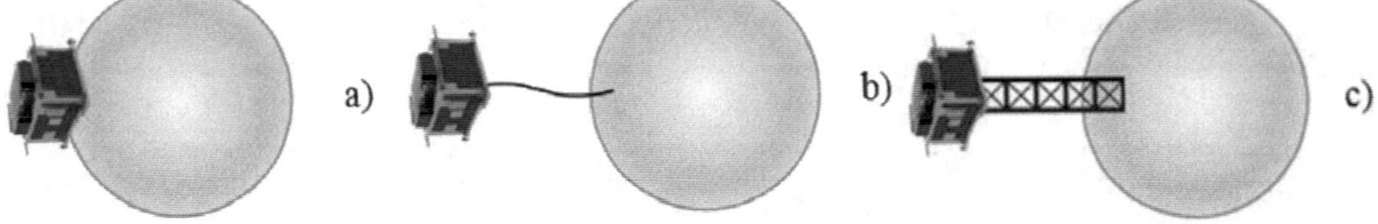

Figure 4. Schemes of mutual placing of satellite and braking device: a) adjacent; b) remote on the cable; c) remote on a hard construction

In this case, it is preferable and technologically feasible to create a braking device shape using inflatable beams (see Fig. 5 and Fig. 6).

That will allow reducing the volumes of working gas for inflation and maintenance of the required form.

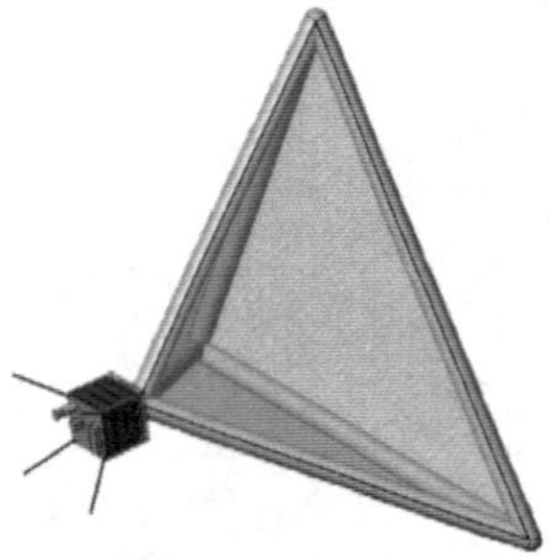

Figure 5. Braking device in the form of a pyramid with ribs in the form of inflatable beams

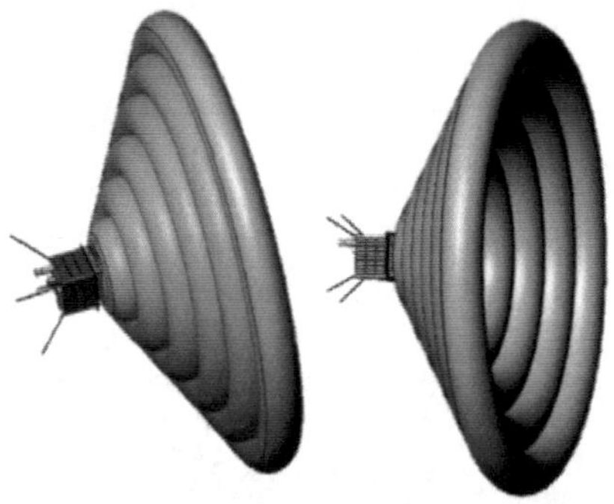

Figure 6. Braking device in the form of a cone with inflatable beams

In addition, one of the tasks is to calculate the volume that will be occupied by the braking device in the folded state.

4. Estimating the lifetime of a spacecraft on orbit

To estimate the spacecraft's orbital lifetime, a software complex for predicting the motion of the center of mass of the spacecraft was developed, which makes it possible to evaluate the possibilities of using braking devices of various configurations.

For an altitude of 200 km, the disturbing effect of atmospheric inhibition is 10^{-4} m/s^2. For an altitude of 400 km, the order of the perturbing acceleration decreases and is 10^{-6} m/s^2.

For altitudes from 800 km to 1500 km, the order of acceleration varies from 10^{-9} m/s^2 to 10^{-12} m/s^2.

However, even a slight perturbing acceleration from the atmosphere on will make a significant contribution to deceleration, since the drag force acts constantly and the direction of action is always opposite to the direction of motion of the space vehicle.

This fact makes it possible to use inflatable, unfolding and other braking devices quite efficiently for the purpose of reducing spacecraft from orbit.

5. Calculation results

As part of the research:

calculation of the rate of fall of the orbit altitude of the spacecraft was carried out for nominal estimates of the forecast of the values of the indices *F10.7* and *ap*;

the mass of satellite varied from 5 kg to 600 kg;

the initial cross-sectional area of the satellite (S0) varied from 0.05 m² to 1 m²;

considered two values of the cross-sectional area of the braking device after opening: $S_1 = 4$ m^2 and $S_2 = 10$ m^2;

the shape of the disclosed braking device was selected as a tetrahedron.

Mass and geometric parameters of satellites with shown in table 1.

IOP Conf. Series: Materials Science and Engineering **812** (2020) 012006 doi:10.1088/1757-899X/812/1/012006

Table 1. Mass and geometric parameters of satellites

Number	Mass, kg	S0, m²
1	5	0.05
2	10	0.10
3	20	0.20
4	30	0.30
5	40	0.40
6	50	0.50
7	100	0.65
8	150	0.75
9	200	0.80
10	300	0.85
11	400	0.95
12	500	0.95
13	600	1.00

The results of calculating the lifetime of satellite on a circular orbit from 300 km to 600 km shown in Fig. 7-11.

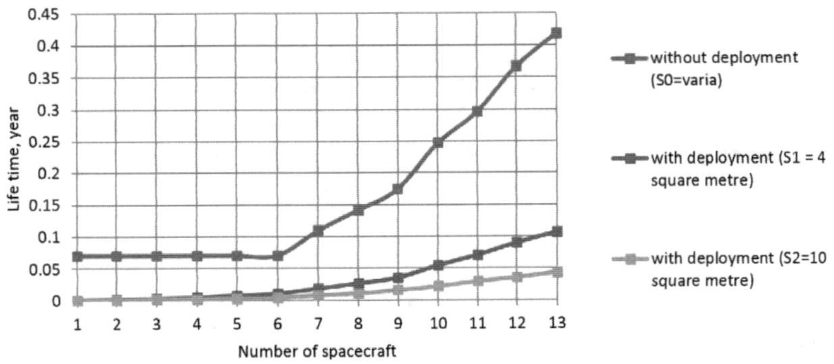

Figure 7. Changes in the height of spacecraft of various configurations for the initial altitude H0 = 300 km

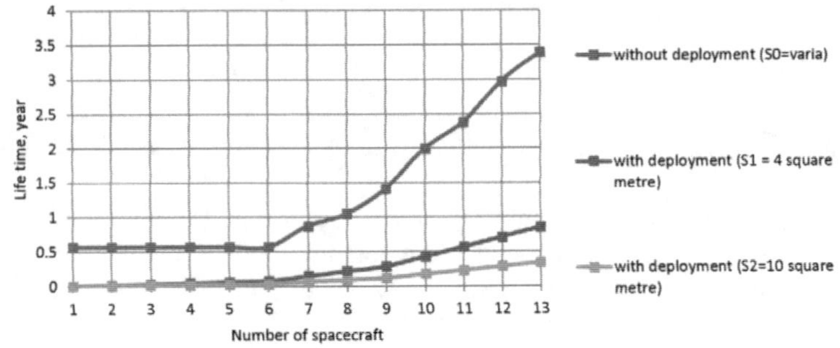

Figure 8. Changes in the height of spacecraft of various configurations for the initial altitude H0 = 400 km

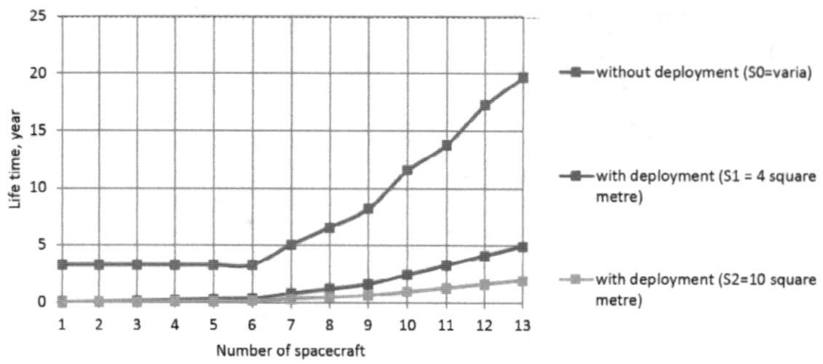

Figure 9. Changes in the height of spacecraft of various configurations for the initial altitude H0 = 500 km

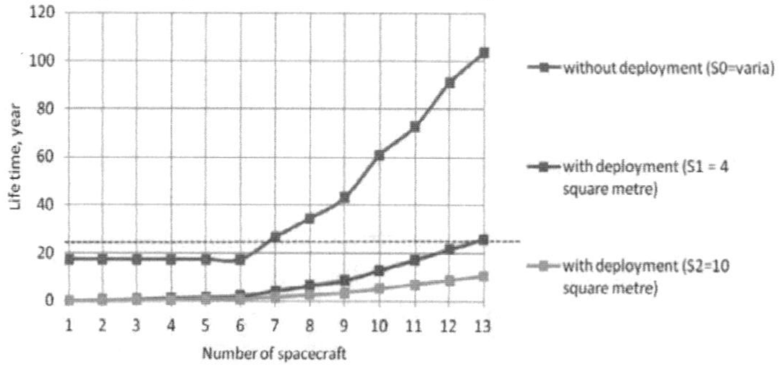

Figure 10. Changes in the height of spacecraft of various configurations for the initial altitude H0 = 600 km

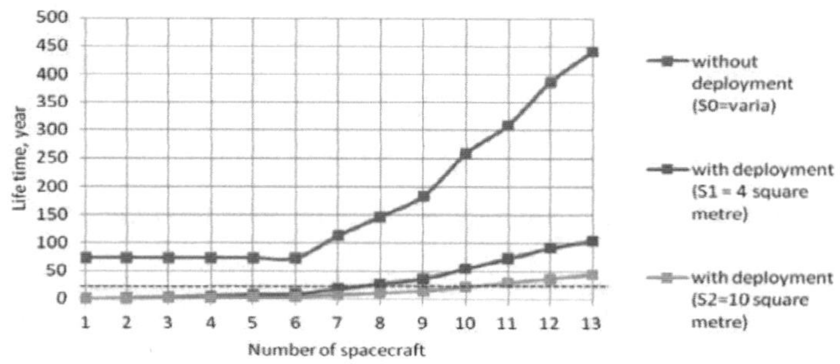

Figure 11. Changes in the height of spacecraft of various configurations for the initial altitude H0 = 700 km

As can be seen from the obtained results Fig. 7-11, the disclosure of the inflatable braking device allows to significantly reduce the duration of passive ballistic spacecraft existence in orbit. And in most cases, the existence time does not exceed 25 years, which meets one of the requirements of the provisions of the Inter-Agency Committee on Space Debris [10].

So, according to preliminary calculations, an inflatable braking device in the form of a tetrahedron with a cross-sectional area of 4 m^2 will weigh about 1.5 kg.

6. Conclusions

Taking into account the obtained results, we can say that the use of passive braking devices in the form of inflatable structures can significantly reduce the time of existence of spacecraft in orbit, so the demand for such systems in the future will be very high.

The main advantage of inflatable structures before rigid ones is a small mass and the possibility of compact stacking in the required volume when putting into orbit. However, to date, inflatable braking devices are not sufficiently tested in outer space conditions, although ground handling has been conducted for a long time.

At the same time, there is a probability of breakdown of the inflatable structure by the fragments of the space debris and the occurrence of leaks of the supercharged gas, which can significantly reduce the efficiency of such a system. It is assumed that such leaks can be compensated by the multiple actuation of the gas generator.

7. References

[1] V.V. Koryanov. Method of calculating the descent of the spacecraft in the atmosphere using technology adaptation landing in different environmental conditions, IAC-14-C2.3.4, 65th Astronautical Congress, Toronto, Canada, 2014, 29 September – **03 October**.

[2] V.V. Koryanov, V.P. Kazakovtsev. Dynamics of angular motion of landing vehicle in martian atmosphere with allowance for small asymmetries. *International Journal of Mechanical Engineering and Robotics Research*, Volume 7, Issue 4, 1 July 2018, Pages 385-391.

[3] Michele Iacovazzo, Valerio Carandente, Raffaele Savino, Gennaro Zuppardi. Longitudinal stability analysis of a suborbital re-entry demonstrator for a deployable capsule. *Acta Astronautica. Volume 106*, January–February 2015, pp. 101–110.

[4] The Mars MetNet Mission. Concept Drawing. http://fmispace.fmi.fi/old-metnet/index.php?id-100, (accessed: 05.02.2019).

[5] Jyri Heilimo, Ari-Matti Harri, Sergey Aleksashkin, Vsevolod Koryanov, Ignacio Arruego, Walter Schmidt, Harri Haukka, Valery Finchenko, Maxim Martynov, Boris Ostresko, Andrey Ponomarenko, Viktor Kazakovtsev, Susanna Martin, and Tero Siili. RITD - Adapting Mars Entry, Descent and Landing System for Earth. Geophysical Research Abstracts Vol. 16, EGU2014-5506-1, 2014, EGU General Assembly 2014

[6] Solar Sail Propulsion [Electronic]. – Access mode: URL: https://ntrs.nasa.gov/archive/nasa/casi.ntrs.nasa.gov/20120015076.pdf. (15.05.2018)

[7] Koryanov V. Research of the dynamics motion of landing vehicle with inflatable braking device in the planet atmosphere // Proceedings of the International Astronautical Congress, IAC vol. 8, 2013. pp. 5831-5836.

[8] Aleksashkin S.N., Pichkhadze K.M., Finchenko V.S. Printsipy proyektirovaniya spuskayemykh v atmosferakh planet apparatov s naduvnymi tormoznymi ustroystvami // Vestnik FGUP NPO im. S.A. Lavochkina [Design principles in planetary atmospheres reentry vehicles with inflatable braking systems // Herald Federal State Unitary Enterprise Scientific and Production Association named after Lavochkin] 2012. №2. pp. 4-11.

[9] V.V. Leonov. The design features of inflatable large-scale mirror concentrators for space high-temperature solar power plant. Proceedings of the international astronautical congress, IAC 67, Making Space Accessible and Affordable to All Countries. 2016.

[10] Report of the Committee on the Peaceful Uses of Outer Space. General Assembly Official Records Sixty-second session Supplement No. 20 (A/62/20). p.50, 2007.

IOP Conf. Series: Materials Science and Engineering **812** (2020) 012007 doi:10.1088/1757-899X/812/1/012007

Structure, luminescence and mechano-optical properties of strontium aluminate doped with europium and dysprosium ions

A. A. Banishev[1] **and A. F. Banishev**[1, *]

[1]Institute on Laser and Information Technologies – Branch of Federal Scientific Research Centre "Crystallography and Photonics" of Russian Academy of Sciences, Svyatoozerskaya Str.1, Shatura 140700, Moscow Region, Russia

[*]Email: banishev@mail.ru

Abstract. Photoluminescence spectra and kinetics of mechanoluminescence of finely dispersed powder of $SrAl_2O_4$:(Eu^{2+}, Dy^{3+}) and $Sr_4Al_{14}O_{25}$:(Eu^{2+}, Dy^{3+}) phosphor was investigated. Photoluminescence was excited by nanosecond laser pulses with a wavelength of λ = 355 nm and by a group of cw lasers with different wavelengths (λ=355, 404, 440, 530 nm). Photoluminescence of the materials has been revealed in the long-wave (550 to 750 μm) spectral region, which is indicative of the existence of corresponding energy levels of Eu^{2+}ions. It was shown that, at low power density, a single wide broadened line of photoluminescence was observed. As radiation power density is increased, additional photoluminescence lines begin to appear. The microstructure of the powders has been investigated with the scanning electron microscope. It has been found that the powder consists of microparticles (granules) of the sizes up to 100 μm. The microparticles show a granular structure, the grain size being as large as 25 μm. Mechanoluminescence was initiated by an acoustic wave generated by short laser pulses. A mechanism of mechanoluminescence excitation has been suggested. It has been concluded that mechanoluminescence of $SrAl_2O_4$:(Eu^{2+}, Dy^{3+}) is excited owing to activation of the traps during their interaction with the electric fields of moving grain boundary dislocations as the result of grain boundary sliding under deformation of microparticles on impact.

1. Introduction

The recent years show the increased interest in the investigation of mechanoluminescence (ML) of materials [1,2]. Mechanoluminescent materials are drawing the increasing attention of specialists engaged in both the fundamental investigations and applied research related to the properties of materials subjected to various force fields. This interest was stimulated, in particular, by the development of a new direction of research – straintronics, which studies the behavior of physical properties and effects in deformed materials. Mechanoluminescence (ML) is a capability of a material to luminesce (glow) as a result of deformation emerging under mechanical stresses. The mechanoluminescent materials are of interest as the basis for creating devices and elements for detecting, monitoring, imaging, and recording mechanical actions (impacts, vibrations, deformations, bending, compression, shock, and etc.) [3-6].

By now, a great quantity of materials is known which effectively transform mechanical actions to optical radiation, as well as to visible radiation. For example, the fine-dispersed powders of $CaAl_2Si_2O_8:Eu^{2+}$ and $SrMg_2(PO_4)_2:Eu^{2+}$ exhibit mechanoluminescence in the blue region [7, 8]; the powders of $SrAl_2O_4:Eu^{2+}$ and $BaSi_2O_2N_2$: Eu^{2+} reveal it in the green region [9-12], and $ZnS:Mn^{2+}$, $CaZnOS:Mn^{2+}$, $Sr_{n-1}Sn_nO_{3n+1}:Sm^{3+}$ – in the red region [13-16]. The work is being continued on the synthesis of new mechanoluminescent materials with higher coefficients of mechanoluminescence efficiency and on studying the mechanisms of mechanoluminescence excitation. The search for new mechanoluminescent materials is continued, including those based on polymer and biopolymer.

The fine powder $SrAl_2O_4$:(Eu^{2+}, Dy^{3+}) attracts attention as a phosphor that is capable of accumulating large light energy and possessing the property of long afterglow. Besides, this material offers the pronounced mechanoluminescence property. In work [18,19] it was established that strong ML of $SrAl_2O_4$:(Eu^{2+}, Dy^{3+}) is only observed in the piezoelectric phase with the twin structure. According to [20,21], the mechanoluminescence of $SrAl_2O_4$:(Eu^{2+}, Dy^{3+}) phosphor powder is attributed to the interaction of the impurity centers with the local piezoelectric fields emerging in the vicinity of the impurity centers under the material deformation. In papers [22-23] the mechanoluminescence of the $SrAl_2O_4$:(Eu^{2+}, Dy^{3+}) phosphor powder is explained by the interaction of the impurity centers with the grain boundary dislocations.

This paper presents the results of studies on photoluminescence spectra and photoluminescence decay kinetics for such luminescent substances as $SrAl_2O_4$:(Eu^{2+}, Dy^{3+}) and $Sr_4Al_{14}O_{25}$:(Eu^{2+}, Dy^{3+}) excited by laser radiation with different wavelengths. The mechanoluminescence was excited by short acoustic pulses generated by laser pulses. The mechanoluminescence (sensor) layer sensitive to mechanical effects was formed directly in the surface layer of a polymer material transparent in the visible region—poly (methyl methacrylate) (substrate).

2. Materials, experimental procedure and results

In the present work, consideration was given to photo- and mechanoluminescence of the fine-dispersed powders of the $SrAl_2O_4$:(Eu^{2+}, Dy^{3+}) and $Sr_4Al_{14}O_{25}$:(Eu^{2+}, Dy^{3+}) phosphors. In particular was studed the spectra and kinetics of photoluminescence of fine-dispersed powders of phosphors $SrAl_2O_4$:(Eu^{2+}, Dy^{3+}) and $Sr_4Al_{14}O_{25}$:(Eu^{2+}, Dy^{3+}). For the excitation of photoluminescence in the samples, were used a set of lasers with different values of radiation wavelength—the third harmonic of a pulsed solid-state YAG:Nd^{3+} laser with λ = 355 nm (τ = 7ns, W = 10 mJ) and semiconductor continuous radiation lasers with $\lambda \approx$ 405 nm, (I = 70 mW).

Figure 1 shows a schematic diagram of the experimental setup. A Clacier-X spectrometer was used to register and analyze the photoluminescence spectra. The photoluminescence signal through an optical fiber was fed to the spectrometer and, with the help of a second optical fiber, was supplied to the input of a monochromator for studying the kinetics of photoluminescence decay.

Figure 2 depicts the photoluminescence spectra of the resulting powders at room temperature. The photoluminescence was excited by the continuous-wave laser with the wavelength λ=405 nm. The photoluminescence of the powders is seen to be a single wide line occupying a large part of the visible spectrum. The wavelengths λ=497 nm and λ=520 nm correspond to the photoluminescence maxima of the $Sr_4Al_{14}O_{25}$:(Eu^{2+}, Dy^{3+}) and $SrAl_2O_4$:(Eu^{2+}, Dy^{3+}) powders, respectively.

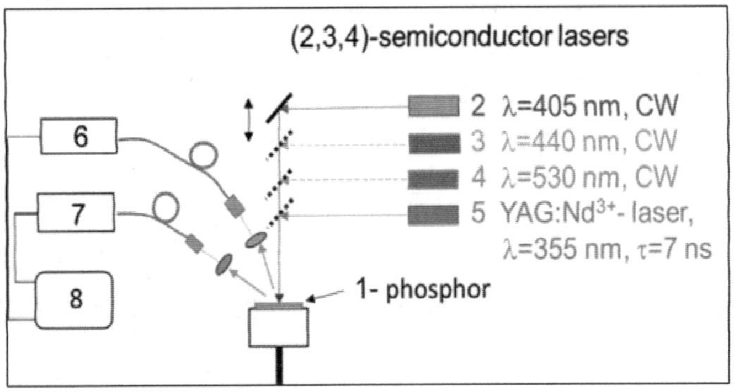

Figure 1. Schematic diagram of experimental setup: (1) luminophore under investigation; (2–5) lasers; (6) spectrophotometer; (7) monochromator; (8) computer.

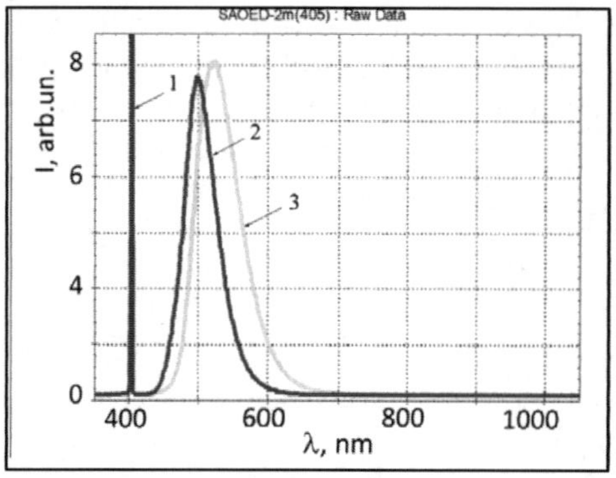

Figure 2. The photoluminescence spectra of the powders under excitation by a laser with λ=405 nm; 1. The radiation of the excitation laser, 2. The photoluminescence spectrum of $Sr_4Al_{14}O_{25}:Eu^{2+}$, 3. The photoluminescence spectrum of $SrAl_2O_4:(Eu^{2+}, Dy^{3+})$/

It is known, that the photoluminescence spectrum of $SrAl_2O_4:(Eu^{2+}, Dy^{3+})$ is determined by the electron transitions of Eu^{2+} ion. The phosphorescence and mechanoluminescence spectra are also governed by the transitions of Eu^{2+} ion, but the intensity and duration of phosphorescence and mechanoluminescence depend to a large extent on the processes of carrier trapping and trap activation. To the traps belong impurities and (or) defects whose energies $E_{n,p}$ are, as a rule, in the forbidden zone near the conduction band bottom or near the valence band top. It may be believe, that Dy^{3+} impurities in the $SrAl_2O_4:(Eu^{2+}, Dy^{3+})$ phosphor form the electron traps.

Figure 3 illustrates the spectra of the $SrAl_2O_4:(Eu^{2+}, Dy^{3+})$ powder photoluminescence excited by the pulses of the laser with λ=355 nm wavelength having various radiation power densities. At low power density, a single wide inhomogeneously broadened line with a maximum at λ=525 nm is observed. As radiation power density is increased, additional photoluminescence lines begin to appear.

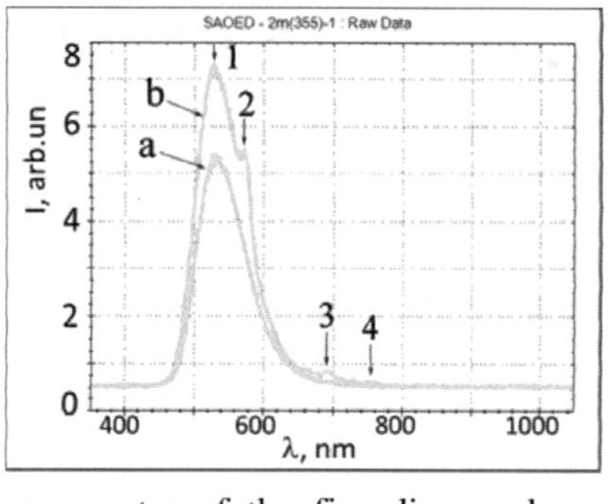

Figure 3. The photoluminescence spectra of the fine-dispersed powder $SrAl_2O_4:(Eu^{2+}, Dy^{3+})$. The laser pulse power density: a) I=5×10^6 W/cm^2; b) I=7×10^6 W/cm^2

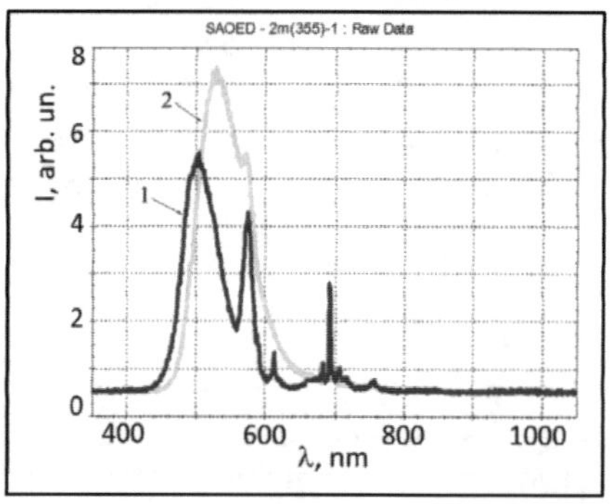

Figure 4. The photoluminescence spectra of the powders under excitation by a laser with λ=355 nm:
1) $Sr_4Al_{14}O_{25}:Eu^{2+},Dy^{3+})$; 2) $SrAl_2O_4:(Eu^{2+}, Dy^{3+})$

Figure 4 shows the photoluminescence spectra of the produced powders under excitation by a laser with the wavelength λ=355 nm and the power density I=7×10^6 W/cm^2. The appearance of the new lines in the photoluminescence spectrum with an increase in the power density testifies that there exist the electron levels and corresponding radiative transitions which have not been registered under the action of laser radiation with low power density.

Figure 5 displays the photos of the microparticles of the $SrAl_2O_4$:(Eu^{2+}, Dy^{3+}) and $Sr_4Al_{14}O_{25}$(Eu^{2+}, Dy^{3+}) powders obtained with the scanning electron microscope. The microparticles are seen to consist of a large number of grains with the distinguished boundaries; the grains are closely adjacent to each other with no pores and microcavities commonly encountered in conglomerates.

In appearance, the microparticles have a polycrystalline structure. The structure of the microparticles and the shape of the grains in $Sr_4Al_{14}O_{25}$:(Eu^{2+}, Dy^{3+}) differ essentially from those of $SrAl_2O_4$:(Eu^{2+}, Dy^{3+}). The grain boundaries are also different in their shape.

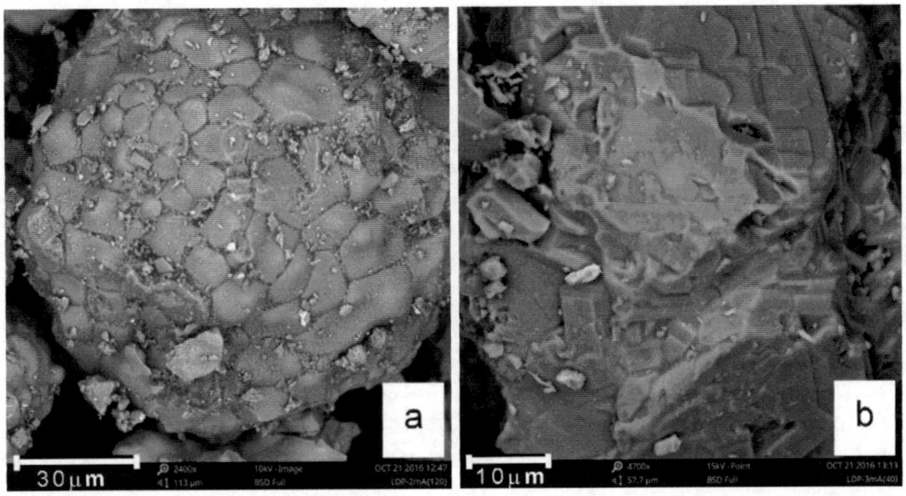

Figure 5. The photographs obtained with the scanning electron microscope: a) A microparticle of the phosphor powder $SrAl_2O_4$:(Eu^{2+}, Dy^{3+}); b) A microparticle of the phosphor powder $Sr_4Al_{14}O_{25}$:Eu^{2+}, Dy^{3+}).

The paper presents the results of investigation mechanoluminescence of the composite materials produced from the photopolymerizing resin (photopolymer) and the fine disperse powder $SrAl_2O_4$:(Eu^{2+},Dy^{3+}). A thin layer (300 µm) of the composite formed on a metal substrate, see figure 6.

Mechanoluminescence was excited by the acoustic pulses emerging under the action of short laser pulses upon the metal plate surface. The laser pulse (λ=355nm, τ=7ns, W=10mJ) was focused to a spot of the diameter d≈1mm (see Figure 6). As the result, a pulse of vapor pressure and acoustic pulse in metal substrate is produced [24].

Figure 7 presents the mechanoluminescence signal of the $SrAl_2O_4$:(Eu^{2+}, Dy^{3+}) powder, excited by an acoustic pulse arising from the action of a short laser pulse. The action of an acoustic pulse brings about the deformation of the polycrystalline microparticles of the phosphor.

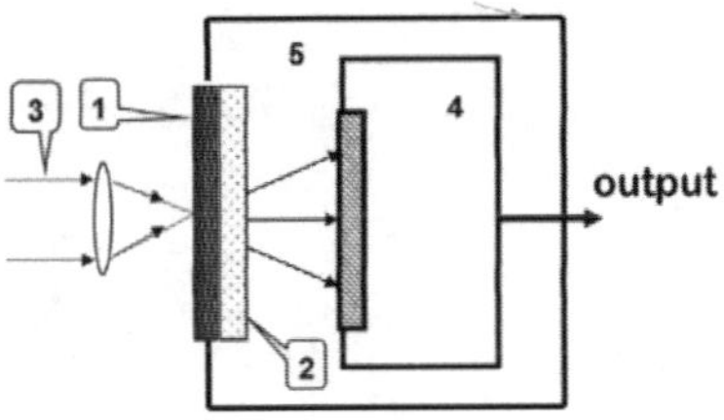

Figure 6. 1)-Substrate (metal plate, h = 100 µm), 2)-Suspension of photopolymer and powder $SrAl_2O_4$:(Eu^{2+}, Dy^{3+}), h = 300 µm, 3)-Laser pulse, 4. Photomultiplier

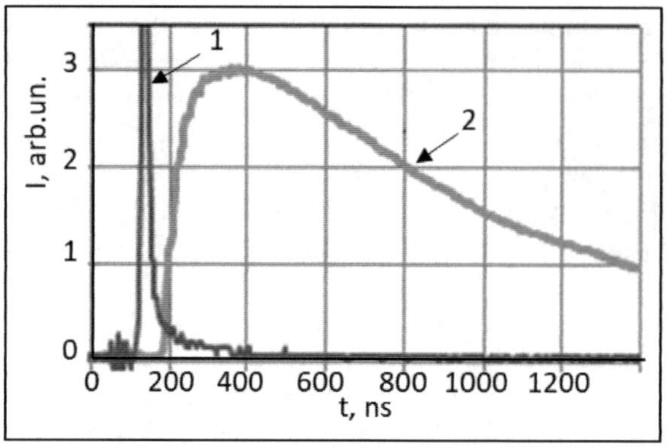

Figure 7. Mechanoluminescence of the $SrAl_2O_4$:(Eu^{2+}, Dy^{3+}) powder excited by a pulse of the laser with W=10 mJ, λ=355 nm.; 1)-Laser pulse, 2)-Mechanoluminescence

In the polycrystalline materials the deformation mainly results from intergranular sliding at the cost of motion of the grain boundary dislocations. It is known, see [25], that there exist rather strong mechanical stresses in the neighborhood of the dislocations, which can cause curving (displacement) of the bands and the levels of impurities and defects. In the stress field of the dislocation, the energy level of a trap is shifted in the direction of the conduction zone. As the result, the energy separation ΔE between the levels of traps and the bottom of the conduction zone is reduced. At the same time, the probability of tunnel transitions of the electrons from the filled trapping levels to the conduction zone is dramatically increased. Then, the nonradiative capture of the electrons from the conduction zone by the Eu^{3+} ions proceeds, leading in appearance of the excited ions Eu^{2+*}. The mechanoluminescence is due to the radiative transitions of Eu^{2+*} ions. It should be noted that $Sr_4Al_{14}O_{25}$:(Eu^{2+}, Dy^{3+}) powder under the same loads does not exhibit any mechanoluminescent properties. The features of mechanoluminescence behavior observed in the experiments agree well with the dislocation mechanism [1, 17] that associates its appearance with curving of the energy bands and levels in the vicinity of dislocations.

3. Conclusion

The paper deals with the mechanism of excitation of the photoluminescence and mechanoluminescence of the cited above powders. It has been shown that the nonuniformly broadened band of photoluminescence of the $SrAl_2O_4$:(Eu^{2+}, Dy^{3+}) microparticles with the maximum λ=525 nm is divided into several spectral bands (573 nm, 693 nm and 755 nm) with an increase in the exciting radiation power. A mechanism of mechanoluminescence excitation is suggested. It is anticipated that mechanoluminescence is excited due to activation of the traps when they interact with the moving grain boundary dislocations at the moment of exposure to the acoustic pulse.

4. Acknowledgment

The work was supported by the Ministry of Science and Higher Education within the State assignment FSRC "Crystallography and Photonics" RAS in part of "production and diagnostics of new nanomaterials and nanostructures for solving topical problems in micro- and nanoelectronics and nanophotonics", Russian Foundation for Basic Research (Project No16-29-14003) in part of "investigation of deformation-stimulated light generation by micro-nanoparticles of phosphor in a matrix of polymer"

5. References

[1] Feng A and Smet P F 2018 *Materials* **11**(4) 41 doi.org/10.3390/ma11040484
[2] Wang X, Peng D, Huangd B, Pan C and Wang Z L *Nano Energy* 2019 **55** 389
[3] Krishnan S, Van der Walt H, Venkatesh V and Vishnu B S 2017 *Journal of Intelligent Material Systems and Structures* **28** (17) 2458

[4] Dong T, Xu C N, Fujio Y and Yoshida A 2015 *Light: Science & Applications* **e356** 1 doi:10.1038/lsa.2015.129
[5] Qian X, Cai Z, Su M, Li F, Fang W, Li Y, Zhou X, Li Q, Feng X, Li W, Hu X, Wang X, Pan C and Song Y 2018 Advanced Materials 30 25
[6] Jha P, Khare A, Singh P K and Chandra V K 2018 *Journal of Luminescence* **195** 40
[7] Zhang L, Yamada H, Imai Y and Xu C N 2008 *J. Electrochem. Soc.* **155** J63
[8] Kamimura S, Yamada H and Xu C N 2012 *J.Lumin.* **132** 526
[9] Rahimi M R, Yun G J, Doll G L and Choi J S 2013 *Opt. Lett.,* **38** 4134
[10] Timilsina S, Lee K H, Jang I Y and Kim J S 2013 *Acta Mater.*, **61** 7197
[11] Banishev A A, Lotin A A and Banishev A F 2014 *Int. J. Modern Phys. B*, **28** 1450154
[12] Banishev A F, Banishev A A and Lotin A A 2012 *Fiz. Khim. Obrab. Mater.* **5** 89 (Russian)
[13] Botterman J, van den Eeckhout K, de Baere I, Poelman D and Smet P F 2012, *Acta Mater.* **60** 5494
[14] Zhang J C, Xu C N, Kamimura S, Terasawa Y and Yamada H 2013 *Opt.Express* **21** 12976
[15] Kamimura S, Yamada H and Xu C N 2012 *Appl. Phys. Lett.* **101**(9) 091113
[16] Bünzli Jean-Claude G, Wong Ka-Leung 2018 *Journal of Rare Earths* **36** 1
[17] Dorenbos P 2005 *J. Electrochem. Soc.* **152** H107
[18] Matsuo H, Ikeda K, Hata S, Nakashima H, Yamada H and Xu C N 2013 Journal of Alloys and Compounds **577** S507
[19] Terasaki N, Yamada H and Xu C N 2013 *Catalysis Today* **201** 203
[20] Jha P and Chandra B P 2014 *The Journal of Biological and Chemical Luminescence, Luminescence* **29** 977
[21] Chandra B P, Chandra V K and Mahobia S K 2012 *Sensors and actuators A: Physicsl* **173** 9
[22] Banishev A A and Banishev A F 2018 *Inorganic Materials: Applied Research* **9** 484
[23] Banishev A A and Banishev A F 2019 Inorganic *Materials: Applied Research* **10** 646
[24] Bunkin F V and Phrochorov A M 1976 *Soviet Physics Uspekhi* **19** (7) 561 http://dx.doi.org 10.1070/PU1976v019n07ABEH005273
[25] Suzuki T, Yoshinaga X and Nakeuchi S 1989 *Dynamics of diclocatins and plasticity* 296p (Russian)

IOP Conf. Series: Materials Science and Engineering **812** (2020) 012008 doi:10.1088/1757-899X/812/1/012008

Finite Element Simulation of Repeated Upsetting at Elevated Temperature

Wei Guo[1, *], Jin Zou[1], Keming Liu[2]

[1] Institute of Applied Physics, Jiangxi Academy of Sciences, Nanchang 330096, China
[2] Jiangxi Key Laboratory for Precision Drive and Control, Nanchang Institute of Technology, Nanchang 330099, China

*E-mail: guowei66@sohu.com

Abstract. The equivalent strain field, effective stress field, and temperature field of the billet after repeated upsetting were analyzed by using finite element simulation software. The results show that the average equivalent strain in the billet improves with the increase of processing pass. At the same pass, the strain in the middle of upper surface is the smallest and in the middle of lower surface is the largest. After three passes, maximum strain of 5.33 is achieved. The equivalent stress and temperature decrease gradually from the middle to both sides of the billet.

1. Introduction

In recent years, the technology of severe plastic deformation has been paid more and more attention by researchers and developed rapidly[1]. There are many factors influencing the effect of severe plastic deformation, such as die structure, processing path, preparing temperature, forming speed, friction coefficient between material and die, etc[2-4]. It takes a lot of manpower and material resource to systematically study the effect of different processing technology on material through experiment. With the rapid development of computer technology and finite element software, it is a shortcut to use numerical simulation to analyze the process of severe plastic deformation by establishing models. The finite element analyses for equal channel angular extrusion, high pressure torsion, accumulative roll bonding and reciprocating extrusion have obtained a lot of useful results[5-8]. Repeated upsetting is a novel severe plastic deformation method, and the remarkable effect of this technology on refining microstructure and improving mechanical properties of Mg alloy has been reported[9-11]. However, the finite element analysis of this technology is still lacking. In this paper, the simulation of repeated upsetting is carried out by using finite element software, and reference is provided for the practical application of this technology.

2. Research procedures

Fig.1a is the sketch map of repeated upsetting process. The device consists of a punch with dimension of $150 \times 100 \times 20$ mm^3 for external loading, a billet with dimension of $100 \times 100 \times 20$ mm^3, and a die with external dimension of $165 \times 180 \times 180$ mm^3. The lower chamber of the die and the billet are of equal dimensions. The simulation of repeated upsetting was carried out by finite element software and three-dimensional models were established. The finite element meshing of the processing device was shown in Fig.1b. The material model of billet was Al-1100 in the database of the software itself, which was defined as a plastic body. The meshing of the model was automatically generated by the software with three-node isoparametric elements. Upsetting process was numerically simulated and the mesh was self-adaptively re-divided. The material of punch and die was H-13 steel, which was defined as rigid body and meshed to facilitate the analysis of heat transfer between punch and billet. For plastic deformation hardly occurred in the punch and die, larger finite element meshes were divide for them. The upsetting temperature of simulation was set to 400 °C. The friction between the billet and die was modeled as shear friction and the friction coefficient was set at 0.3. The downward speed of the punch was 4 mm/s during upsetting, and three processing passes were made. After the first pass was simulated, the billet was taken out and rotated 90° around the X-axis and replaced into the die cavity. Then the simulation results were imported into the pretreatment module of the next upsetting pass and the second pass was carried out. In the same way, the third upsetting pass was completed. In the post-processing module, the values and distributions of strain, stress, and temperature in the billet after upsetting were obtained.

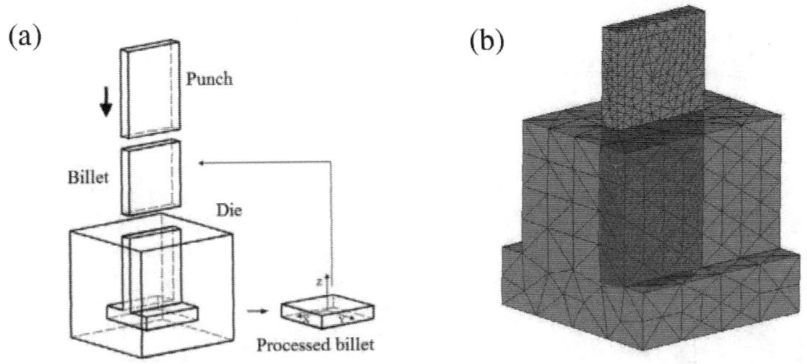

Figure 1. (a) Sketch map and (b) finite element mesh generation of repeated upsetting device

3. Results and analyses

3.1. Strain field

The strain distribution in the billet after repeated upsetting for one, two and three passes are shown in Fig.2a, Fig.2b and Fig.2c, respectively. The strain presents gradient distribution from the middle to both sides. High strain distributes in the middle position of the bottom, due to this part deforms from the beginning to the end of upsetting. With the increase of passes, the maximum equivalent strain in the billet increases from 2.37 to 5.33. The low strain mainly distributes in the middle part of upper surface, for this part begins to deform just before the end of upsetting. With the increase of passes, the minimum equivalent strain value increases from 0.708 to 3.49. The increase of the strain value indicates that the microstructure of the billet will be gradually uniformed and refined. In addition, the side of the billet changes from flat to a certain radian, the edges and corners become more and more obscure with the processing pass.

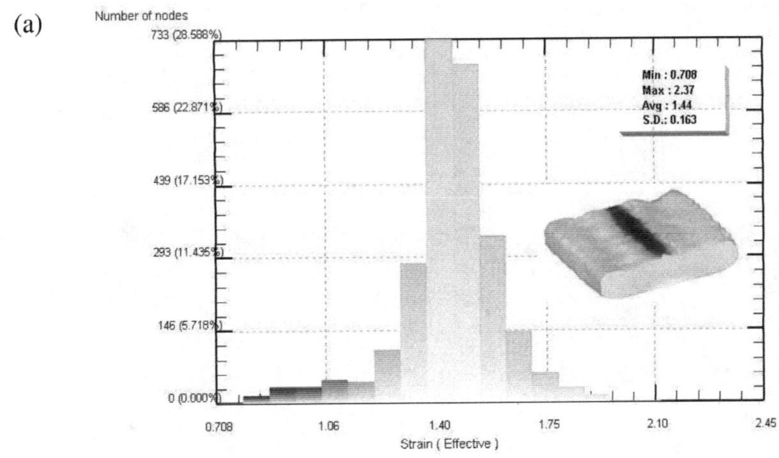

(b)

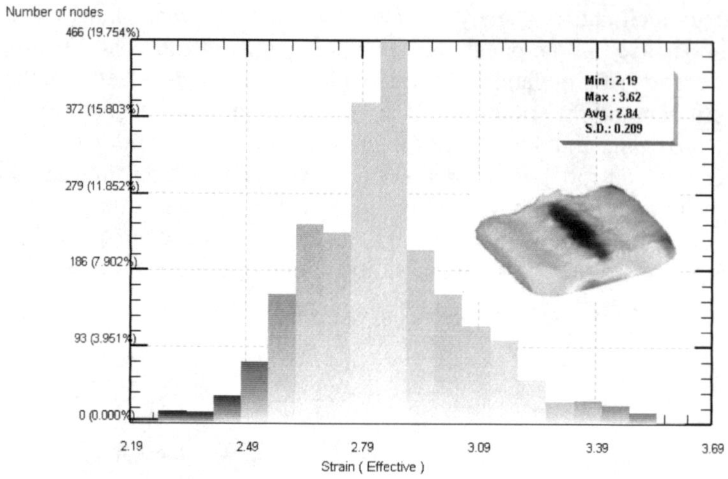

(c)

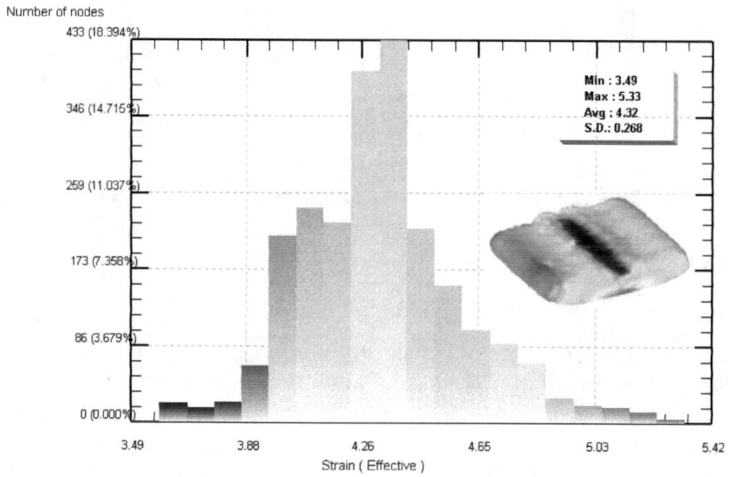

Figure 2. The strain distribution in the billet after repeated upsetting for (a) one, (b) two, and (c) three pass

3.2. Stress field

As shown in Fig.3, the effective stress distribution in the billet after one pass upsetting is between 10.8 MPa and 45.1 MPa. The highest stress distributes at the edge contact between the billet and the punch, which is directly affected by the die at the corner. The medium-sized stress alternately distributes in the billet from the middle part to both sides.

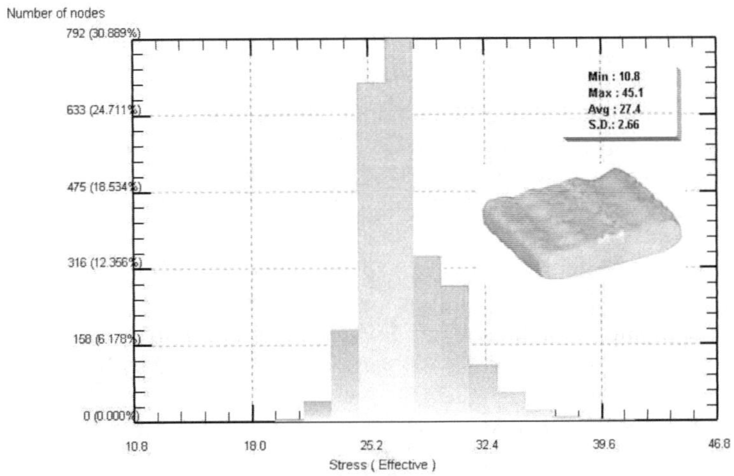

Figure 3. Distribution of effective stress (MPa) in the billet after one pass upsetting

3.3. Temperature field
The temperature distribution in the billet after one pass upsetting is shown in Fig. 4. It can be seen that the temperature presents a remarkable gradient distribution. The temperature of the punch contact area is the highest, reaching 411 °C, due to the heat generated by the friction between the billet and the die. The temperature decreases gradually along the edge of the billet, and the lowest temperature at the corner is 404 °C.

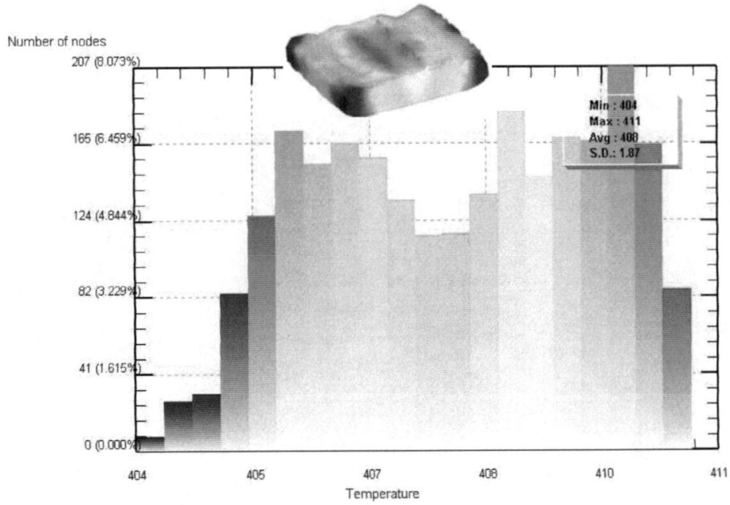

Figure 4. Temperature (°C) distribution in the billet after one pass upsetting

4. Summary
In this paper, the equivalent strain, effective stress, and temperature field of billet during repeated upsetting were analyzed by using finite element simulation technology. The main conclusions are as follows:

(1) The average equivalent strain in the billet increases with the increase of upsetting pass, and the strain in the upper middle part is the smallest, while that in the bottom middle part is the largest.

(2) The high stress zone in the billet is the edge contact area between the billet and the punch.

(3) Temperature in the billet presents a gradient distribution, which gradually decreases from the punch contact area to the edge, and the lowest temperature distributes at the corner.

5. Acknowledgments
The present research was sponsored by the Key Scientific Research Project from Jiangxi Academy of Sciences under Grant nos. 2017-YZD2-20, the National Natural Science Foundation of China (NSFC) under Grant No. 51861025, the Provincial Natural Science Foundation of Jiangxi under Grant No. 20192BAB206001, Scientific Research and Development Foundation from Jiangxi Academy of Sciences under Grant nos. 2018-YCXY-11 and 2018-YYB-08.

6. References
[1] Valiev R Z, Islamgaliev R K, Alexandrov I V, *Prog. Mater. Sci.* Vol. **45** (2000) pp. 103-189.
[2] Nie K B, Wu K, Wang X J, et al, *Mater. Sci. Eng. A* Vol. **527** (2010) pp. 7364-7368.
[3] Zhao Z D, Chen Q, Tang Z J, et al, *J Alloys Comp.* Vol. **497** (2010) pp. 402-411.
[4] Gutierrez-Urrutia I, Mu˜noz-Morris M A, Puertas I, et al, *Mater. Sci. Eng. A* Vol. **475** (2008) pp. 268-278.
[5] Wei W, Zhang W, Wei K X, et al, *Mater. Sci. Eng. A* Vol. **516** (2009) pp. 111-118.
[6] Yoon S C, Horita Z, Kim H S, *J. Mater. Process. Technol.* Vol. **201** (2008) pp. 32-36.
[7] Inoue T, Yanagida A, Yanagimoto J, *Mater. Lett.* Vol. **106** (2013) pp. 37-40.
[8] Lin J B, Wang Q D, Peng L M, et al, *J Mater. Sci.* Vol. **43** (2008) pp. 6920-6924.
[9] Guo W, Wang Q D, Ye B, et al, *Mater. Sci. Eng. A* Vol. **540** (2012) pp. 115-122.
[10] Liu J F, Wang Q D, Zhou H, et al, *J Alloys Comp.* Vol. **589** (2014) pp. 372-377.
[11] Zhou H, Ye B, Wang Q D, et al, *Mater. Lett.* Vol. **83** (2012) pp. 175-178.

IOP Conf. Series: Materials Science and Engineering **812** (2020) 012009 doi:10.1088/1757-899X/812/1/012009

The Integration technology of desulfurization, denitration and heat recovery from coke oven flue gas

Hui Jiang[1], Shun Li[2], Jun Ha [3],Wei Guo[2] , Yixin Pan[2], Xue Wan [2] ,Guowei Xie[2]

[1]Zhongwei National Engineering Research Center For Coking Technology Co., Ltd. , Anshan, Liaoning, 114044, China
[2]Sinosteel Anshan Research Institute of Thermo-energy Co., LTD., Anshan, Liaoning, 114044, China
[3]Anshan Iron and Steel Group Corporation, Anshan, Liaoning, 114021, China

E-mail: asrnls2008@126.com

Abstract: The main components of coke oven flue gas were SO_2, NO_X etc. SO_2 and NO_X were not only harmful to human health, but also seriously polluted the environment. The stricter "Standard for Discharge of Pollutants from coking chemical industry" was introduced. An integration settlement scheme of flue gas desulfurization, denitration and waste heat recovery was put forward. The proprietary core technologies, such as denitration, COG desulfurization and radial heat pipe boiler were researched. The integration settlement scheme can solve the problems of polluting the environment and waste heat without recovery, which had not only significant economic benefits, but also huge social benefits.

1. Introduction

Coking plant was a special plant of production coke and processing coke products. Flue gas of coke oven was exhaust gas after combustion, whose harmful gas components were mainly SO_2, NO_X etc. Pollutants had the characteristic of organization and continuous emission, so coking industry was one of the more serious pollution of industry.

New "Standard for Discharge of Pollutants from coking chemical industry" clearly defined standards for the discharge of atmospheric pollutants in coking industry. With the very stringent standard and the increasingly tight energy supply more advanced, economic, reasonable coke oven flue gas processing method was needed.

2. Integration settlement scheme

A new integration settlement scheme of flue gas desulfurization, denitration and waste heat recovery was put forward. The flue gas treatment process was Coke Oven Gas desulfurization, combustion, flue gas denitration, radial heat pipe boiler, booster fan and chimney. Its core technologies were denitration and radial heat pipe boiler.

2.1.COG desulfurization before combustion

COG desulfurization methods in large coking plants were mainly wet desulfurization. The principle was the reaction of desulfurization catalyst and H_2S in alkaline solution[1]. Elemental sulfur was precipitated with air regeneration. Desulfurization before chemicals and desulfurization after chemicals were two main methods of wet desulfurization in the coking plant. The former was that COG from the condensation blast firstly processed desulfurization with ammonia, and followed chemicals[2]. The method eliminated the corrosion of H_2S on subsequent equipment, but generally H_2S concentration decreased to 200 mg/m^3,which need additional second desulfurization equipment to achieve delivery standard.

The latter was that COG from the condensation blast firstly processed chemicals with Na_2CO_3, and followed desulfurization[3]. The method had heavy corrosive on subsequent equipments and need at the same time consumed a large amount of Na_2CO_3 in order to guarantee the alkalinity index. But the method had high desulfurization, up to the requirements of 20mg/m^3, without additional second desulfurization device.

2.2.Denitrification reactor

At present the common denitration methods [4～5] were selective non catalytic reduction (SNCR), oxidation absorption method and selective catalytic reduction (SCR) etc. Selective catalytic reduction (SCR) was currently the most mature and efficient method for exhaust gas denitrification. The principle of SCR for NOx removal was that on catalyst surface NOx was reduced to N_2 with ammonia added in exhaust gas, the reaction equation follows.

$$4NOx+4NH_3+O_2 \rightarrow 4N_2+6H_2O$$

Ammonia water or gas containing a certain concentration of ammonia was introduced to denitrification reaction with pipe. It flowed into the mixer and uniformly mixed with flue gas. The sensors were respectively arranged at the reactor inlet and outlet for on-line real time monitoring NOx concentration controlling the amount of ammonia according to the feedback signal.

The denitration catalyst was very important in the denitrification process. A new denitration catalyst was developed. The catalyst used the whole coating structure with ceramic honeycomb, which was composed by the ceramic honeycomb, metal oxide coating and active component. Metal oxide coating uniformly and firmly attached to the outer surface of the ceramic honeycomb, then active component was dispersed on the oxide coating. The catalyst was a very suitable monolithic catalyst for flue gas denitration, which had advantages of high airspeed operation, low resistance, good selectivity, low ammonia leak rate, wide temperature range and small thermal expansion coefficient.

When NOx concentrations was very high at the entrance high accuracy NOx removal also can achieve, and even the concentration of NOx can be less than 150mg/m^3. If the emission standards is further upgraded, the catalyst and reaction device does not have to alter. If the amount of ammonia slightly increases, then the NOx concentration in the exhaust gas is less than the specified limit.

Denitration industry test was done in a coking plant, whose experimental device was shown in Figure 1.

Figure 1. Denitration industry test device

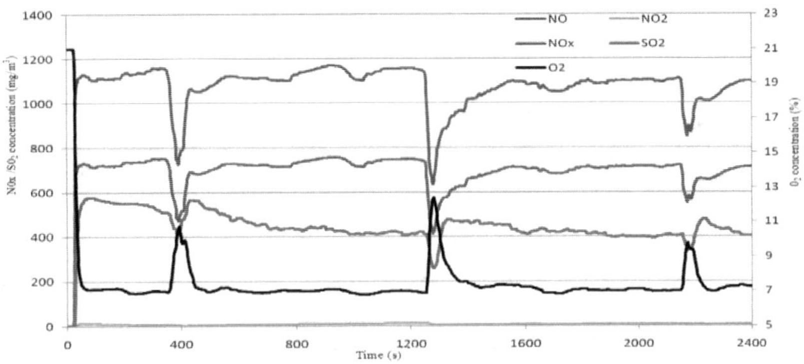

Figure 2. Gas components before denitration

Without denitration treatment, continuous monitoring concentration of NOx, NO_2, SO_2, O_2 and NO respectively was 1100 mg/m^3, 720 mg/m^3, 450 mg/m^3, 190 mg/m^3 and 10 mg/m^3 or so at the outlet of butterfly valve, the cyclical fluctuations of gas composition every 15 minutes was produced by coke oven operation system, as shown in Figure 2.

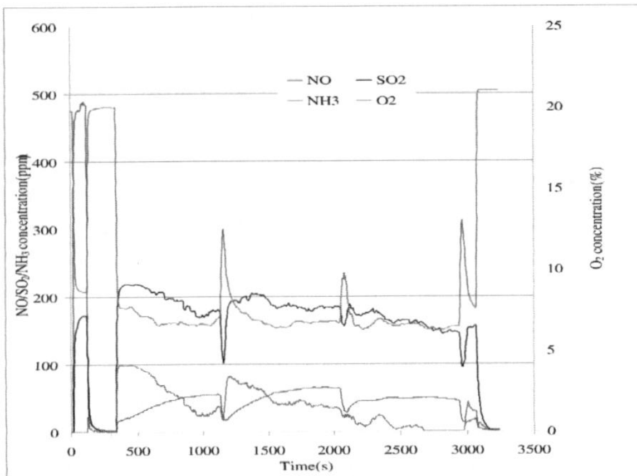

Figure 3. Gas components after denitration

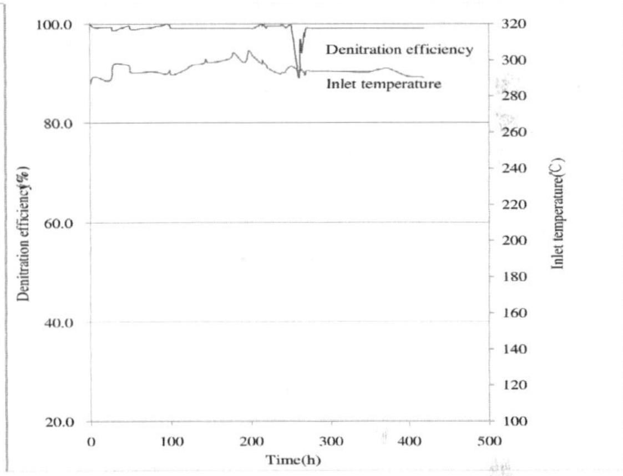

Figure 4. Denitration efficiency

From Figure 3 we can see when a certain amount of ammonia was pumped, NOx concentration was 30ppm, which was accord with the national standards. Up to now, the experiment has been running for almost 1000 hours, the following conclusions were obtained from comprehensive denitrification experimental results.

- The denitration efficiency reached stablely more than 99% (measured NO content was several ppm, which can be understood as systematic error), closed to 100%. The denitrification efficiency of the catalyst was super high, as shown in Figure 4;
- The experimental operation airspeed was about 16000 h-1, which was 4 times that of traditional denitration catalyst;
- The pressure drop in catalyst bed was about 100 Pa and can significantly reduce the energy consumption of the fan;
- Modular design of catalyst can avoid farthest future engineering amplification problem and was beneficial to the realization of engineering amplification.

2.3. Radial heat pipe heat-exchanger

The conventional axial heat pipe was applied to heat recovery from coke oven flue gas. It had advantages of efficient superconductor equipment, excellent isothermal homeothermia, excellent unidirectional thermal-transfer characteristics, good environmental adaptability and avoid dew point corrosion, but disadvantages of complex structure, huge equipment, easy tube explosion, damageing irreversibility and relatively high equipment cost.

IOP Conf. Series: Materials Science and Engineering **812** (2020) 012009 doi:10.1088/1757-899X/812/1/012009

In view of the advantages and disadvantages of the axial and heat pipe a new radial heat pipe heat-exchanger was designed.

●The eccentric radial heat pipe heat-exchanger had some advantages of eccentricity scientifically designed and heat pipe working medium increased to achieve a reasonable state of working medium, improve the thermal efficiency of heat pipe heat-exchanger, reduce overall project cost and raise the production efficiency, whose energy-saving efficiency reached the international advanced level.

●According to the existing problem of the axial heat pipe heat-exchanger at home and abroad, the technology effectively resolved problems such as complex structure, big steel consumption to steam production, irreversible loss and low thermal efficiency after damaging for the axial heat pipe heat-exchanger.

Eccentric radial heat pipe developed was successfully applied to 4.3m coke oven in a domestic company coking plant. The key equipment operation was stable and steam production capacity can meet the requirements, air had not blocking phenomenon, whose pressure control was good in the bottom of the furnace flue.

Per ton coke can produce saturated steam 68.8 kg, namely 0.19 GJ/t, in the same period the radial heat pipe consumed only 1.29kW/t.

3. Conclusions

Denitration catalyst had advantages of high efficiency and operation airspeed, low pressure drop. Modular design of catalyst can avoid farthest future engineering amplification problem and was beneficial to the realization of engineering amplification.

New radial heat pipe heat-exchanger effectively resolved problems such as complex structure, big steel consumption to steam production, irreversible loss and low thermal efficiency after damaging for the axial heat pipe heat-exchanger.

The scheme can solve the problems of polluting the environment and waste heat without recovery, which had not only significant economic benefits, but also huge social benefits.

4. References

[1] YANG Li, ZHANG Li-Ying. Coke Oven Gas Desulphurization and Decyanation Methods, J.HEBEI HUAGONG, 34(2011):5-9.

[2] WU Bing-Zhi, KUAI Zheng-Hao. Determination of Sulfur Content in Iron Minerals by High Frequency Infrared Absorption Spectrometer, J.Technology & Development of Chemical Industry, 40(2011):31-33.

[3] JIANG Wei. The Compar ison of Various Desulphurizing Methods of Coke- oven Gas, J.SCI-TECH INFORMATION DEVELOPMENT & ECONOMY, 17(2007):278-279.

[4] WU A-Feng, LI Ming-Wei, HUANG Tao and TAN Can-Shen. Several Practical Flue Gas de-NOx Technologies and Techno-economic Analysis, J.ELECTRIC POWER, 39(2006):71-75.

[5] MA Dong-Zhu, ZHANG Ling, LI Shu-Shan and YIN Di. Application and Development of SCR Denitration Technology in Coal-fired Power Plant, J.Coal Technology, 30(2011):709-713.

Structure load calculation for Essential ESW Intake Structure /CCW HX Building Chilled Water Compression Tank

Muhammad Sajjad[1] and Dong Won Jung[2*]

[1,2]Department of Mechanical Engineering, Jeju National University, Jeju, South Korea

[*]E-mail: jdwcheju@jejunu.ac.kr

Abstract. Interest in studying of the shell arises from the fifties of twentieth century. Assemblies which contains thin shells, find wide use in the modern engineering especially in ships, aircraft and spacecraft industry. In today's aerospace and aircraft industries, structural efficiency is the main concern. In this paper the shell vibrations and buckling modes are analyzed by means of numerical methods to clarify qualitatively critical loads and different buckling modes. In our study we focus on the experimental analysis of axial loading for static and buckling analysis for pressure vessel. The modeling and the analysis both are carried out in Ansys 15.0 solver. The analysis provides the stresses for the saddle support & shell near the saddle support due to upset (OBE) and faulted (SSE) seismic load conditions. Furthermore, the results from the simulation are then compared with the KEPIC (Korea Electric Power Industry Code) standards to find the structure integrity of particular unit. Finally, data collected from these simulations results, we conclude that it is safe to manufacture particular pressure vessel with current loading conditions.

1. Introduction

Pressure vessels are important because many liquids and gases are stored in it under high pressure [1]. Special emphasis is placed upon the strength of the vessel to prevent explosions as a result of rupture. Codes for the safety of such vessels have been developed that specify the design of the container for specified conditions. Most pressure vessels are required to carry only low pressures and thus are constructed of tubes and sheets rolled to form cylinders [2]. Some pressure vessels must carry high pressures; however the thickness of the vessel walls must increase in order to provide adequate strength [3]. Interest in studying of the shell arises from fifties to twentieth century. The assemblies containing thin shells find wide use in the modern engineering especially in ships aircraft and spacecraft industry. The basic schematics of pressure vessel is shown in Figure 1.

Cylindrical shells such as thin-walled laminated composite unstiffened vessels like deep submarine exploration housings and autonomous underwater vehicles are subjected to combination of in plane, out of plane and shear loads due to the high external hydrostatic pressure during their application [4,5]. Due to the geometry of these structures, buckling is one of the most important failure criteria [6,7]. Buckling failure mode of a stiffened cylindrical shell can further subdivided into global buckling, local skin buckling and stiffener crippling. Global buckling is collapse of the whole structure, i.e. collapse of the stiffeners and the shell as one unit. Local skin buckling and the stiffeners crippling on the other hand are localized failure modes involving local failure of only the skin in the first case and the stiffeners in the second case [8].

IOP Conf. Series: Materials Science and Engineering **812** (2020) 012010 doi:10.1088/1757-899X/812/1/012010

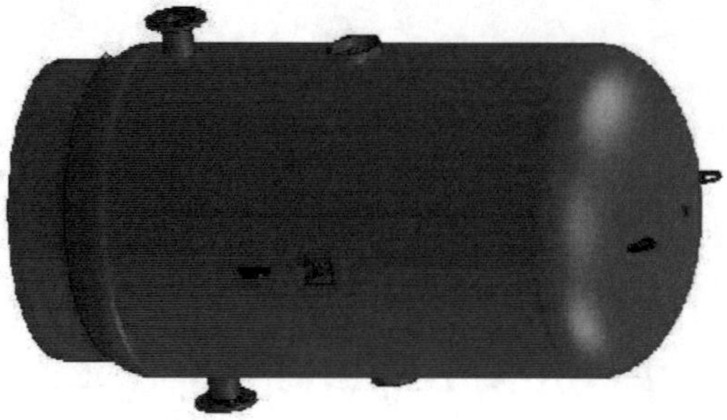

Figure 1. Chilled Water Compression Tank

2. Problem identification

In today's aerospace and aircraft industries structural efficiency is the main concern. Due to their high specific strength and light weight, fiber reinforced composites find a wide range of applications [9]. Compression load carrying structures form part of all aircraft and space vehicle fuel tanks, air cylinders are some of the many applications. An analytical procedure is developed to design and predict the behaviour of Essential ESW Intake Structure/ CCW HX Building Chilled Water Compression Tank. Stress analysis has been carried out for Saddle supports and vessel shell near the saddle support and their effects on Chilled Water Compression Tank pressure vessel has been studied [2]. The shell vibrations and buckling modes are analysed by means of numerical methods, to clarify qualitatively the critical loads and different buckling modes.

3. Modelling of cylindrical pressure vessel

For the design data the weight of empty tank is consider 570 kg which is full of water the total weight is 1070 kg and the pressure is 80 pascals while the operating temperature is 65.5 C^{o}. The modelling of the Cylindrical Pressure Vessel is done in ANSYS 12.1.

There are two type of seismic loads applied in study. One is equivalent static analysis used for the design of anchor bolt and BSLS (Building Structure Load Summary) calculation [10,11]. The other is dynamic analysis for the stress calculation of saddle support and shell near the saddle support due to seismic loads including sloshing effects. The following data shall be applied for the dynamic analysis.

SSE seismic response spectra for 3% of damping for SSE and 1/2 SSE seismic response spectra for 2% of damping for OBE from Technical Specification. The load combinations for Equipment & supports shall be in accordance with the Table 1. The following loading combination and stress limits shall be applied for the Building Structure Load Summary and structure stress analysis in this paper.

Table 1. Stress limits for Equipment & supports

Condition	Loading Combination	Loading Combination	Strength Limits
Normal	W +P + T + O	(0.9)W+(1.7)P+(1.3)T+(1.4)O	
Upset	W +P + T + O + S(u)	(0.9)W+(1.7)P+(1.3)T+(1.4)O+(1.9)S(u)	ACI 349-01
Faulted	W +P + T + O + S(y)	(0.9)W+(1.0)P+(1.0)T+(1.0)O+S(y)	

Where W is Weight (operating weight), P is Pressure Load, T is Thermal Expansion, O is Equipment Operating Loads (Piping Loads), S(u) is Upset condition Dynamic Loads (operating basis earthquake load : 1/2 SSE with 2% damping) and S(y) is Faulted condition Dynamic Loads (safe shutdown earthquake load : SSE with 3% damping).

4. Finite element model

The finite element analysis code for this work is the ANSYS 12.1 ANSYS shell elements type SHELL181 (4-noded structural shell) were selected for modeling. Figure 2 is finite element model for analysis including applied boundary conditions. All nodes on the anchor chair in the skirt support were restrained vertically and laterally. ANSYS 'POST1' will categorize stress into membrane, bending, membrane + bending, peak, and total stresses at inside & outside surface and center of cut section.

Since the ANSYS program uses linear distribution of stress through the thickness of shell elements, intensity values of shell mid-surface, shell bottom and top surface present linearized results. The mid-wall stress is a shell membrane stress while the top and bottom surfaces are membrane plus bending stresses. Hence, the mid-wall stress can be compared directly with the KEPIC code for allowable stress of membrane stress and the two-surface membrane plus bending stresses.

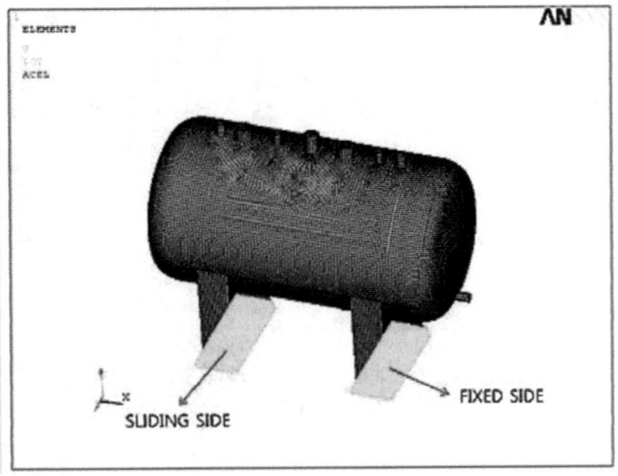

Figure 2. Boundary conditions applied finite element model

5. Building structure load calculation and modal analysis

Building Structure Load calculation shall be summarized for the Operating load, Nozzle loads and seismic loads. Equivalent static seismic loads, which is 1.5 times maximum acceleration in the response spectra curves is applied for the anchor bolt design and the calculation of BSLS (Building Structure Load Summary) conservatively. The arrangement of anchor bolt and its dimensions are shown in Figure 3.

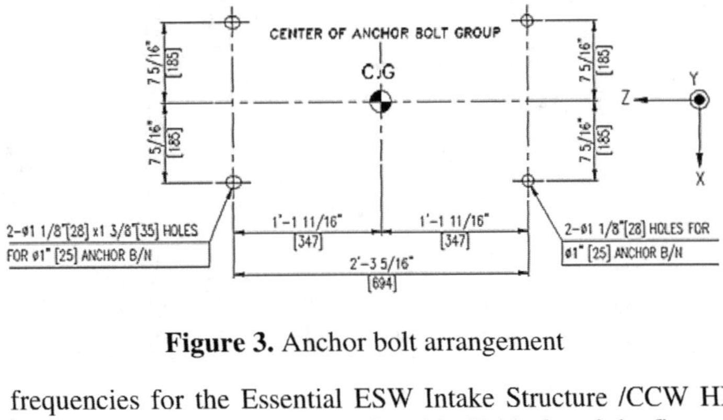

Figure 3. Anchor bolt arrangement

The five modal frequencies for the Essential ESW Intake Structure /CCW HX Building Chilled Water Compression Tank in cycles per second are listed in Table 2 and the first and 2th mode shapes (Figure 4 are attached).

Table 2 Modal frequencies at different mode

Mode sequence number	Frequency (Hz)
1	53.2
2	59.0
3	123.3
4	127.8
5	131.4

1st Mode – f_n = 53.2 Hz

2nd Mode – f_n = 59.0Hz

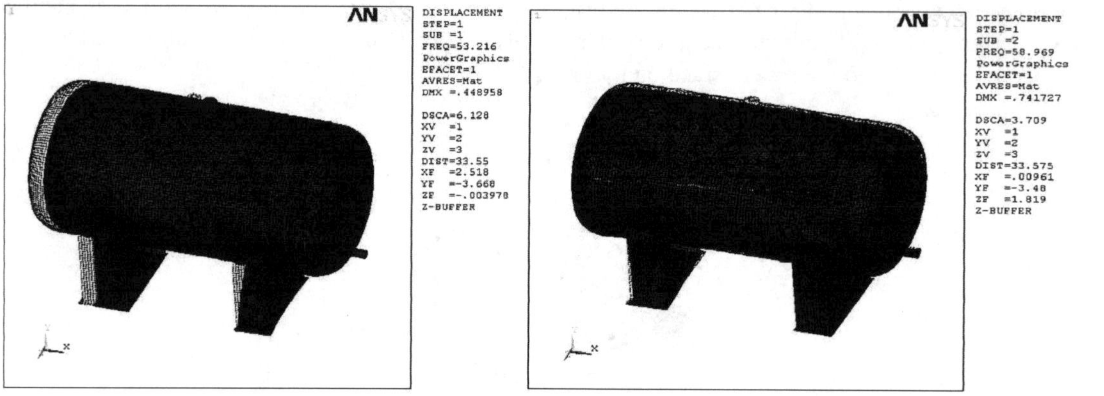

Figure 4. Mode shapes and natural frequencies of 1st and 2nd Mode

6. Seismic loading

The CCW HX Building Chilled Water Compression Tank behaves like as rigid body because the lowest fundamental frequency is higher than 33 Hz. Therefore, constant accelerations are applied to three directions as shown in Table 3.

Table 3 Direction of frequency for spectrum curve and FE model

Spectrum Value (g)			Applied Value (g)		
Direction in spectrum curve	SSE	OBE	Direction in FE model	SSE	OBE
EW	0.35	0.175	X & Y	0.40	0.20
NS	0.40	0.2	X & Y	0.40	0.20
VT	0.37	0.185	Z	0.37	0.185

7. Stress analysis

ANSYS is used for static, piping, seismic OBE and SSE in the load combination and combined stress shall be checked with the applicable stress limits for the following areas.

1) Saddle supports.
2) Vessel Shell near the saddle support

ANSYS output shows three kinds of stresses, which are stresses inside, outside and middle of each element. Stress in middle is membrane stress while stress those are inside and outside of vessel are called bending stress.

7.1. Saddle supports

The saddle supports are SA36, S = 14.5 ksi, Sy = 36 ksi, Su=58 ksi, the maximum stress intensity of saddle support is indicated through Table 4 and Figure 5,6 and 7.

IOP Conf. Series: Materials Science and Engineering **812** (2020) 012010 doi:10.1088/1757-899X/812/1/012010

Table 4 Maximum Stress Intensity of Saddle supports

Load	Operating condition	Classification	Max. Stress Intensity	Refer to	
Static load	Normal	Pm	1.29	Fig.5	
		Pm + Pb	4.05		
	Upset	Pm	1.29		
		Pm + Pb	4.05		
	Faulted	Pm	1.29		
		Pm + Pb	4.05		
Piping load	Normal	Pm	10.22	Fig.6	
		Pm + Pb	15.28		
	Upset	Pm	10.22		
		Pm + Pb	15.28		
	Faulted	Pm	16.18		
		Pm + Pb	24.14		
Seismic load	Normal	Pm	N/A	Fig. 7	
		Pm + Pb	N/A		
	Upset	Pm	0.70		
		Pm + Pb	1.06		
	Faulted	Pm	1.40		
		Pm + Pb	2.12		
Total	Normal	Pm	11.51	< F_{allow} = 14.5 ksi	OK
		Pm + Pb	19.33	< 1.5F_{allow} = 21.75 ksi	
	Upset	Pm	12.21	< F_{allow} = 19.29 ksi	OK
		Pm + Pb	20.39	< 1.5F_{allow} = 28.94 ksi	
	Faulted	Pm	18.87	< F_{allow} = 40.6 ksi	OK
		Pm + Pb	30.31	< 1.5F_{allow} = 60.9 ksi	

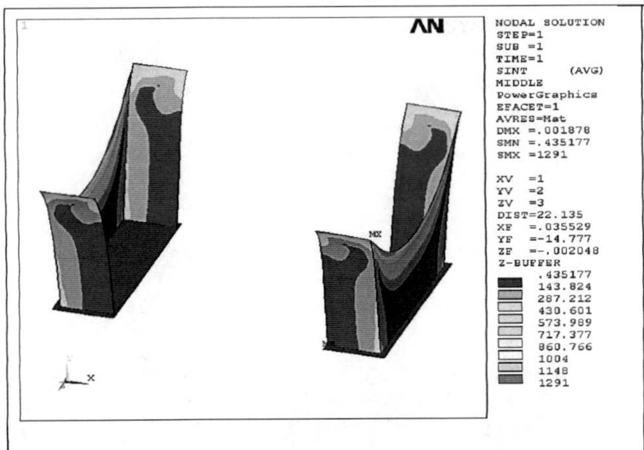

Figure 5. Distribution of Membrane stress intensity of saddle support (Static load)

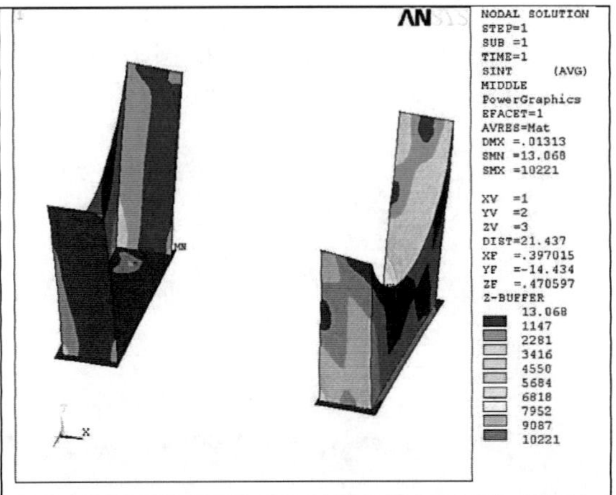

Figure 6. Distribution of Membrane stress intensity of saddle support (Piping load - Normal /Upset)

IOP Conf. Series: Materials Science and Engineering **812** (2020) 012010 doi:10.1088/1757-899X/812/1/012010

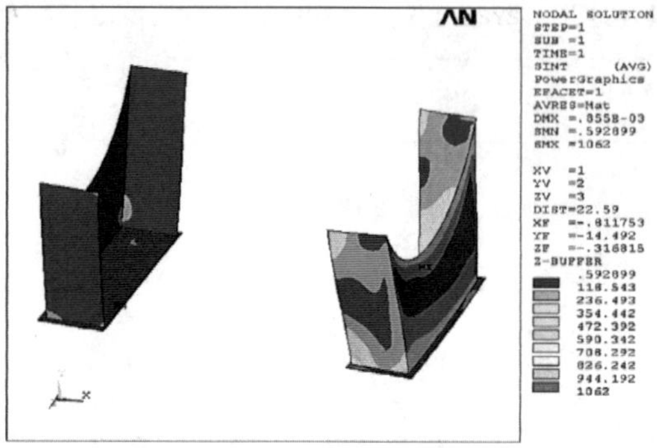

Figure 7. Distribution of Membrane+ Bending stress intensity of saddle support (Seismic load - SSE)

7.2. Vessel shell near the saddle support

The vessel shell stresses are calculated by ANSYS for the internal pressure, weight and SSE and piping loads. The piping stresses is combined using the square root of sum of squares (SRSS) method in the ANSYS program and added to the sum of pressure, weight and 1/2 SEE seismic event stresses for the upset condition. The piping stresses are combined using the SRSS method in the ANSYS program and added to the sum of pressure, weight and SSE seismic event stresses for the faulted condition.

The maximum principal stresses in the shell occur at the intersection of the piping and are largely due to the piping loads. The stresses in the supports and in the shell at the support pads are not as large. All the data is clearly shown in Table 5 and Figure 8, 9 and 10.

Table 5 Maximum Stress Intensity of Shell near the saddle supports

Load	Operating Condition	Classification	Max. Stress Intensity	Refer To
Static Load	Normal	Pm	2.44	FIG. 8
		Pm + Pb	4.07	
	Upset	Pm	2.44	
		Pm + Pb	4.07	
	Faulted	Pm	2.44	
		Pm + Pb	4.07	
Piping Load	Normal	Pm	3.09	FIG.9
		Pm + Pb	6.70	
	Upset	Pm	3.09	
		Pm + Pb	6.70	
	Faulted	Pm	5.02	
		Pm + Pb	10.86	
Seismic Load	Normal	Pm	N/A	FIG. 10
		Pm + Pb	N/A	
	Upset	Pm	0.16	
		Pm + Pb	0.37	
	Faulted	Pm	0.32	
		Pm + Pb	0.74	
Total	Normal	PM	5.53	

Unit : ksi

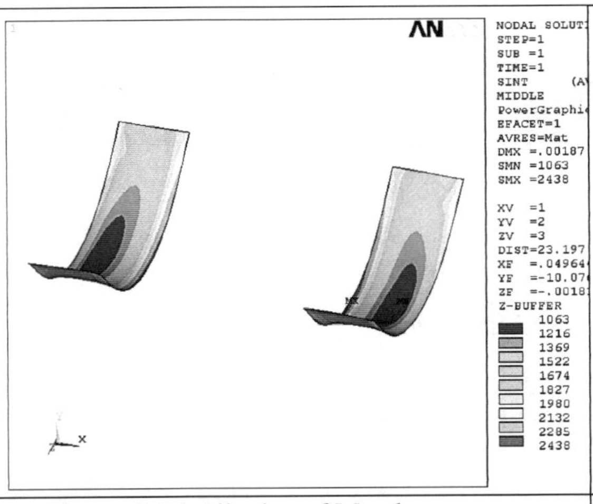

Figure 8. Distribution of Membrane stress intensity of shell near the saddle (Static load)

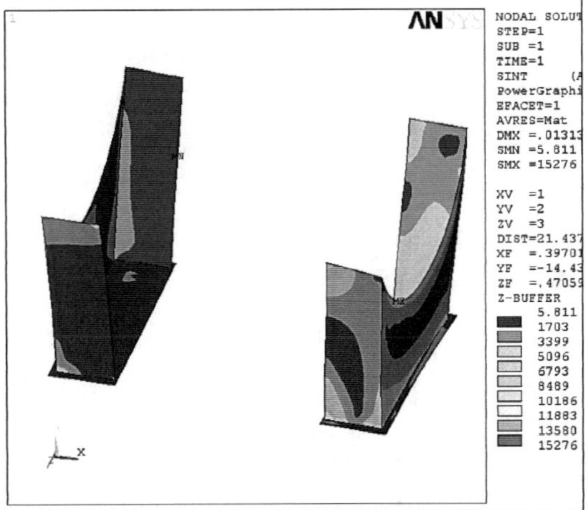

Figure 9. Distribution of Membrane+ Bending stress intensity of saddle support (Piping load - Normal /Upset)

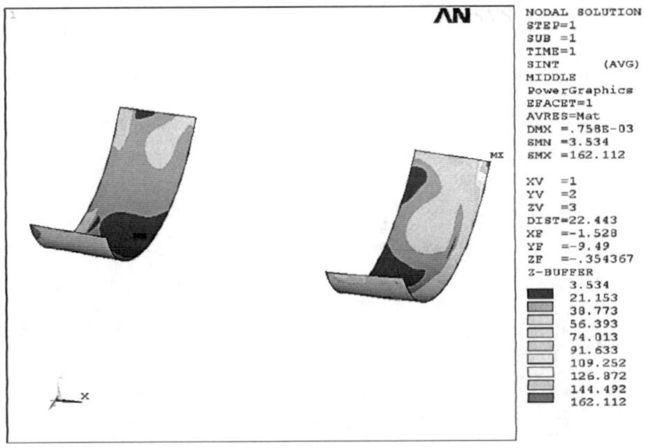

Figure 10. Distribution of Membrane+ Bending stress intensity of shell near the saddle (Seismic load - OBE)

8. Allowable KEPIC stresses

According to KEPIC MDP, the allowable stress limits of components for Normal, upset and Faulted condition for primary Su plus secondary Sy stresses. The comparison between the resulting stresses and allowable stress limits are summarized in Table 6. According to this standard the ratio of maximum stress divide by allowable stress should be less than 1 for the structural integrity of Compression Tank during operation.

Table 6 Stress Evaluation Results of KEPIC standards

Stress	Material	Su (ksi)	Sy (ksi)
Shell, Head	SA-516-70	70	38
Flange	SA-105	70	36
Nozzle	SA-106-B	60	35
Fitting	SA-105	70	36
Base Plate	SA-36	58	36
Anchor bolt	A-193-B7	125	105

IOP Conf. Series: Materials Science and Engineering **812** (2020) 012010 doi:10.1088/1757-899X/812/1/012010

9. Maximum stress intensities due to Hydrostatic test (MNC 3218)

Our model is subjected to each direction force and moment imposed simultaneously on the equipment nozzles, seismic loads and static loads. The analysis provides the stresses for the saddle support & shell near the saddle support due to upset (OBE) and faulted (SSE) seismic conditions. Actual test pressure shall not exceed the required test pressure by more than 6% as shown in table 7.

Sy = 38.0 ksi
Pm < 0.9 Sy = 34.2 ksi
Pm +Pb < 1.35 Sy = 51.3 ksi   For Pm < 0.67 Sy = 25.46 ksi
Pm + Pb < (2.15Sy-1.2Pm)   For 0.67 Sy = 25.46 ksi < Pm < 0.9 Sy = 34.2 ksi

Table 7 Stress values for different load conditions

Load	Classification	Max. Stress Intensity	Refer to
Hydrostatic test	Pm	10.82	Fig. 11,12
	Pm + Pb	15.13	

Pm = 10.82 ksi   < Fallow = 34.2 ksi, ratio of Pm/Fallow<1   OK
Pm + Pb = 15.213ksi   < Fallow = 51.3 ksi, ratio of Pm+Pb/Fallow<1   OK

Since the ratio of (Pm/Fallow, Pm+Pb/Fallow) is less than 1.0, it is concluded that the structural integrity of Essential ESW Intake Structure CCW HX Building Chilled Water Compression Tank is maintained during operation as shown in Figure 11 and 12.

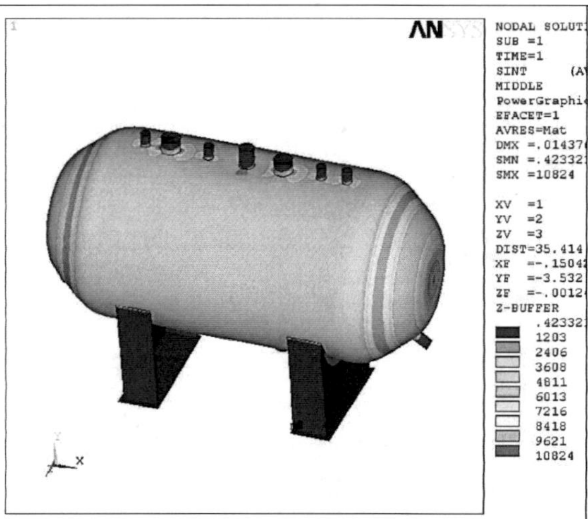

Figure 11. Distribution of Membrane + Bending stress intensity (Hydrostatic test)

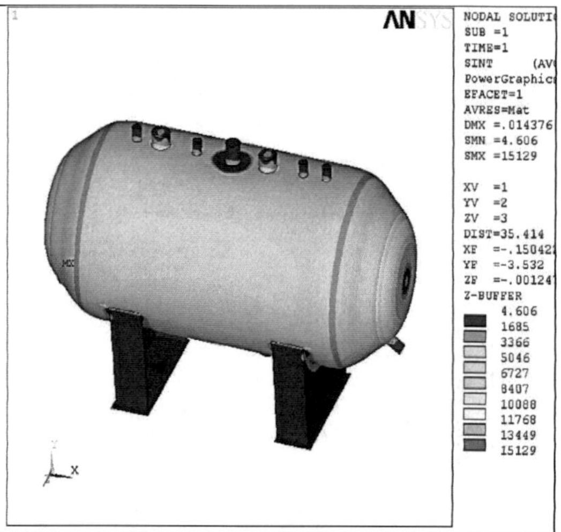

Figure 12. Distribution of Membrane + Bending stress intensity (Hydrostatic test)

10. Conclusion

This study represents the results of a Dynamic Analysis (Seismic) of the proposed Essential ESW Intake Structure CCW HX Building Chilled Water Compression Tank. Dynamic analysis was performed using ANSYS computer program and the structure was modeled using finite element techniques.

The model was subjected to each direction force and moment imposed simultaneously on the equipment nozzles, seismic loads and static loads. The analysis provides the stresses for the saddle support & shell near the saddle support due to upset (OBE) and faulted (SSE) seismic conditions. The seismic analysis is performed based on the lowest fundamental frequency (53.2 Hz) of ESW Compression Tank, therefore it is considered as rigid. The constant seismic acceleration values which is shown in in Technical specification shall be applied for equivalent static analysis instead of spectrum analysis. The partial components for the x, y, and z directions were combined using the square root of the sum of the squares (SRSS) method. Finally, the results from the experiment are compare with KEPIC standards to find the structure integrity of the pressure vessel. Form the result we conclude that it is safe to design the particular unit for given amount of static load, piping load and Seismic Loads.

11. References

[1] Rao Yarrapragada K.S. S R.Krishna Mohan B.Vijay Kiran, 2012 Composite Pressure Vessels *International Journal of Research in Engineering and Technology* vol 01

[2] Graham J. 1993 preliminary Analysis Techniques for Ring and Stringer Stiffened Cylindrical Shells *NASA report TM-108399*

[3] Tae-Uk Kim. Hvo-Chol Sin. 2001 Optimal design of composite laminated plates with the discreteness in ply angles and uncertain in material properties considered (*Computers and Structures* vol 79) pp 2501-2509

[4] Mouritz. A.P. Gellert. E. Burchill. P and Challis. K. 1996 Review of advance composite structures for naval ship and submarines (*Composite structures* vol 53) pp129-139

[5] Alexis A. Krikanov. 1999 Composite pressure vessels with higher stiffness" Elsevier science *Ltd. Composite structures* vol 48) pp 119-127

[6] S. Adali, E.B. Summers & V.E. Verijenko. Optimization of Laminated Cylindrical pressure vessels under strength criterion (*Composite structures* vol 25) pp 305-312

[7] Park O. Haftka R.T. Sankar V.V. and Nagendra S. 1993 Analytical and Experimental Study of a blade stiffened panel in Axial Compression *proceedings of the39th AIAA/ ASME/ ASCE/ AHS/ ASC Structures Structural Dynamics and Materials Conference* pp 20-23

[8] Jaunky N. Knight N.F. Ambur D.R.,1998 Optimum Design and General Stiffened Composite Circular Cylinders for Global Buckling with Strength Constraints (*Journal of Composite Structures*)

[9] Levend Parnas. Nuran Katirei. 2002 Design of Fiber-Reinforced Composite pressure vessels under various loading conditions (*Composite structures* vol 58) pp 83-95(2002).

[10] S.B. Filippov D.N. Ivanov N.V. Naumova. 2005 Free Vibrations and Buckling of a Thin Cylindrical shell of Variable Thickness with Curvilinear Edge

[11] Hyer M.W. Riddick J.C. 2001 Effect of Imperfections of Buckling and Post buckling Response of Segmented Circular Composite Cylinders *Proceedings of the 6th Annual Technical* Conference *of the American Society for Composites Virginia*

IOP Conf. Series: Materials Science and Engineering **812** (2020) 012011 doi:10.1088/1757-899X/812/1/012011

Research and modeling of wrinkles and control of rectangular membrane structures with high-class modeling in on-orbit conditions

S Kumar[1,a], S H Upadhyay [2,b], V V Koryanov[3,c]

[1]BIT Sindri, Dhanbad-828123, India
[2]Smart Materials and Structures Laboratory, Department of Mechanical & Industrial Engineering Indian Institute of Technology, Roorkee-247667, India
[3]Dynamics and flight control of rockets and spacecraft department Bauman Moscow State Technical University, Moscow, Russia, 105005

Email: [a] skumarnitj@gmail.com, [b] shumefme@iitr.ac.in, [c] vkoryanov@bmstu.ru

Abstract. In the previous paper, the authors described a mathematical model of the process, this paper is devoted to the issues of modeling and analysis. The effects of different parameters such as temperature gradients, membrane's aspect ratio, pre-stress, thickness, anchor points, and elastic properties of the membrane and the multilayer on the characteristic wrinkling behavior (wavelength, size, critical load and location) have been examined. It is found that the critical buckling condition of the membrane is dependent on the compressibility, anchor points, aspect ratio and the thickness of the membrane, and it's out of plane displacement depends on the magnitude of the applied in-plane tension. The number of wrinkles decreased with increasing the number of anchor points of the membrane. An effective active flatness control method is used to minimize the surface error (RMS error). Also, it is seen that the multilayer geometry helps in reducing the surface error of the membrane structure.

1. Introduction

The large space structures are required for the deep space exploration, high speed communication, and accurate earth observation. But the weight of the payload increases when the size of structures increase in the conventional space structures and probability of the successes decreases. Also, the operation cost required to transport mass to the destination is high as well as the volume of payload is limited. Hence, achieving minimal launch mass and volume of space structures is always an important task for the space mission. [1]–[5]. Therefore, ultra-lightweight and ultra-large membrane structures are the key element of inflatable reflector, solar sail, solar collector, sun-shields, deployable mirrors surface for the next generation of spacecraft. Many such membrane based space structures have been tested like IN-Step Inflatable antenna experiment [6], [7], IKAROS [8], CubeSat [9], [10].

Thin membranes are often subjected to stresses and strains when an external force applied. Membrane structure is very flexible and has strong geometric nonlinearity and can leads to an out of plane displacement due to their negligible flexural rigidity and presence of in-plane compressive stresses [2], [11]–[14]. Hence it is essential and desirable for a better understanding of the effects of wrinkles on the structural performance and its stability. Surface wrinkling may also be utilized to develop techniques for probing the surface characteristics of materials. [13], [15], [16].

Many authors have discussed the various analytical, numerical and experimental methods to study the wrinkling behavior of thin membrane structures [16]–[25]. Reissner (1938) derived the tension field theory without initial tension under in-plane torsion to and analyzed the circular membrane [26]. Stein and Hedgepeth (1961) to analyze partly wrinkled membranes using Stein's theory [27]. Using the Rayleigh-Ritz method, the circular stretched membrane is analyzed without bending rigidity, but geometrical stiffness considered by [28]. Damil et al [29] have discussed a wrinkling phenomenon using a new multiscale approach taking one-dimensional beam model. Blandino et al [30] have discussed the wrinkling phenomena at the isothermal condition and prove that number of wrinkles increase with the tensile loads. Also, the amplitude of wrinkles increases with increase in thermal gradient. The full shell methods and the reduced resolution methods can be used to solve the wrinkle problems. For the thermo-mechanical load, Kodjo et al [31] have modified Roddeman's model [18] to considering thermal effect in the model.

In this paper, thermal effect on flat membrane antenna has been discussed at orbital conditions. Also, authors discussed the wrinkling analysis of the membrane structures under different loading conditions, the effect of prestress, thickness variation, elastic properties, anchor points. Thermal stresses have been introduced in the classical equation of within the framework of the elastic shell theory. Here Asymptotic Numerical Method is used to solve the nonlinear equations and effective active flatness control method is used to minimize the RMS error of the surface wrinkles of the membrane structure. Also, the analysis of bi-layer structure has been performed to minimize the wrinkled area.

2. Wrinkles Details

The deformed shape under these unsymmetrical loading is shown in Fig. 1, and it is significantly different from symmetric loading a continuous, large diagonal wrinkle goes between the two more heavily loaded corners. In addition to this diagonal wrinkle, fans of small wrinkles can still be seen near the other two corners. The out-of-plane displacements in this plot have been amplified 100 times, for clarity of wrinkle details. The comparison of out of plane displacement for all T_1/T_2 ratios are shown in Fig. 2 at distance 0.280 m from the corner.

3. RMS Error Calculation

The RMS error of wrinkle amplitude has been calculated by.

$$RMS\ Error = \sqrt{\frac{\Sigma(\bar{x}-x)^2}{n}} \quad (1)$$

Where $\bar{x}$ the mean value of amplitude and n is the total number of nodes on the membrane structure. First RMS error has been calculated for all cases (T_1/T_2=1 to 5) and plotted in with respect to T_1/T_2 ratios as shown in Fig. 3. Graph clearly shows that the RMS error increases with increases in the load ratios (T_1/T_2).

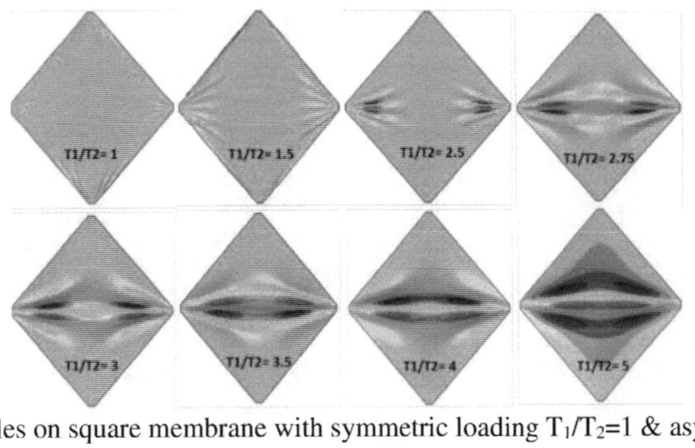

Figure 1. Wrinkles on square membrane with symmetric loading T_1/T_2=1 & asymmetric loading ($T_1/T_2\neq1$)

IOP Conf. Series: Materials Science and Engineering **812** (2020) 012011 doi:10.1088/1757-899X/812/1/012011

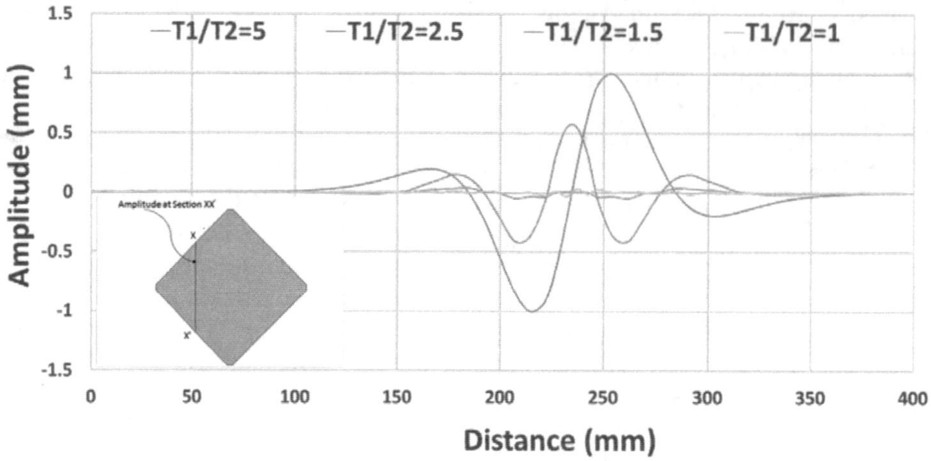

Figure 2. Out of plane displacement with respect to different T_1/T_2 ratio of 0.5 m x 0.5 m square membrane

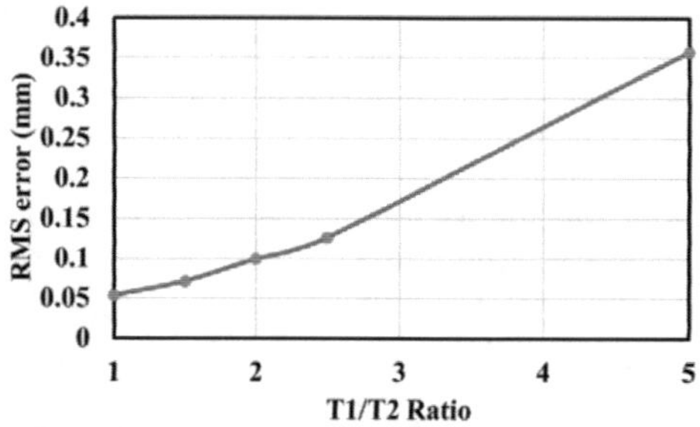

Figure 3. Comparison of RMS Error with respect to T1/T2 ratios

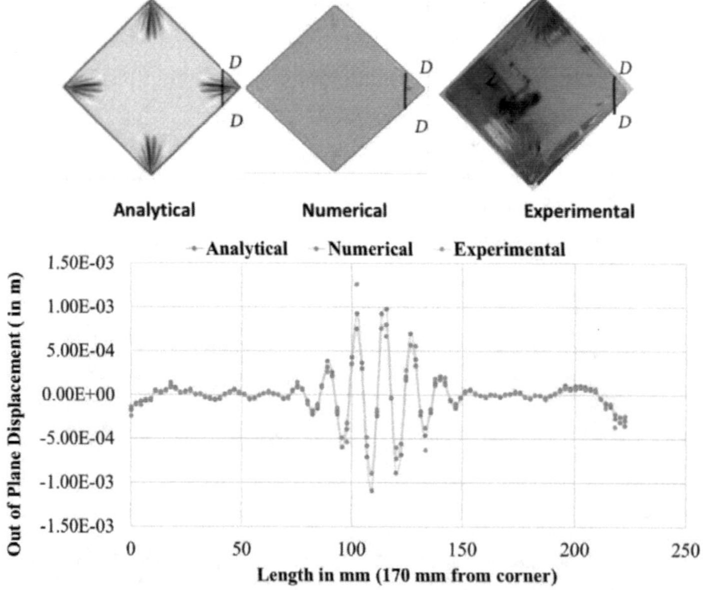

Figure 4. Comparison of Analytical, Numerical and Experimental Results of the 0.5 m x 0.5 m membrane

4. Comparison of Numerical and Experimental Results of Square Membrane

In this sub-section, numerical and experimental are compared with under various loading conditions the wrinkles details. Fig. 4 shows the comparison of wrinkles amplitude (out–of-plane displacement) of 0.5 m x 0.5 m Aluminized Kapton membrane under corner tensile loading at 170 mm rom the corner. Experimental plot confirms the validation of numerical and analytical model. The RMS error of same is listed in Table 1. Similarly, Out-of-plane displacement of the 1.0 m x 1.0 m square membrane is compared at 250 mm from the corner for $T_1/T_2=1$ as shown in Fig 5. At Fig. 6 shows the comparison of winkle's amplitude along midway of the membrane side for $T_1/T_2=5$. The comparison of experiment and numerical value of amplitude of wrinkles at 170 mm from the corner of the Kapton membrane with size of 0.5 m x 0.5 m under different T_1/T_2 ratios are shown in Fig. 7. The RMS error of different size of membrane with various T_1/T_2 ratios are compared in the Table 2. Previously, it confirmed that RMS error increase with increase in T_1/T_2 ratios. RMS error increase with increase in size of membrane.

Table 1. RMS error comparison of 0.5 mx 0.5 m membrane

	RMS Error (mm)
Analytical	0.052
Numerical	0.053
Experimental	0.056

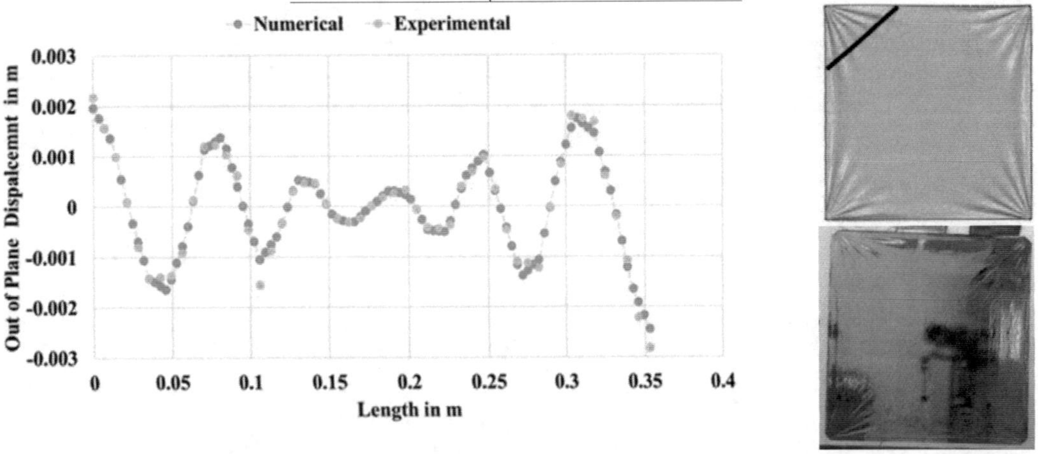

Figure 5. Comparison of experimental and numerical results of out of plane displacement at a distance 250 mm from corner of the 1.0 mx 1.0 m square membrane with symmetric loading (T1/T2=1)

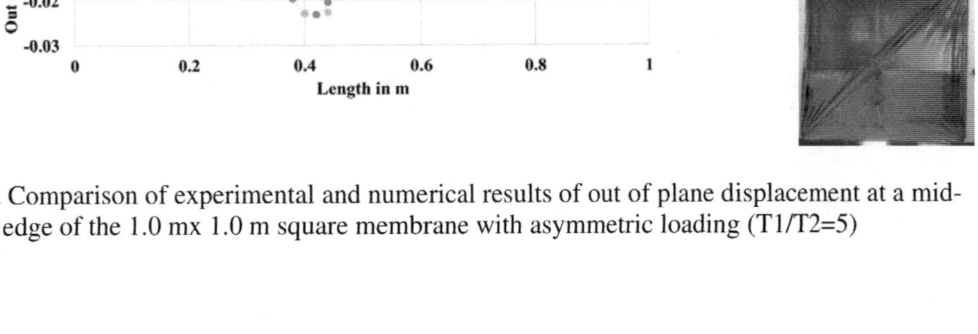

Figure 6. Comparison of experimental and numerical results of out of plane displacement at a mid-edge of the 1.0 mx 1.0 m square membrane with asymmetric loading (T1/T2=5)

IOP Conf. Series: Materials Science and Engineering **812** (2020) 012011 doi:10.1088/1757-899X/812/1/012011

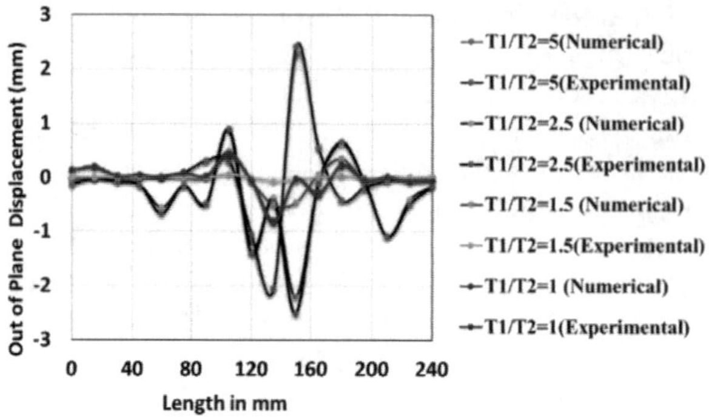

Figure 7. The comparison of experiment and numerical value of amplitude of wrinkles at 170 mm from the corner of the membrane (size of 0.5 m x 0.5 m)

Table 2. RMS error of different size of Square Membrane with various T1/T2 ratio

T1/T2	RMS error (mm)					
	0.2 m²	**0.25m²**	**0.5 m²**	**0.75m²**	**1.0 m²**	**1.2 m²**
1	0.035	0.038	0.053	0.079	0.091	0.156
1.5	0.058	0.063	0.071	0.113	0.143	0.298
2	0.072	0.076	0.099	0.178	0.253	0.589
2.5	0.095	0.113	0.126	0.257	0.509	0.883
5	0.253	0.293	0.357	0.659	0.996	1.256

5. Effect of Pre-Stretching and Thickness

This sub-section considers the variation of the wrinkle details with the magnitude of the applied loads. Fig. 8 compares the out of plane displacement at cross-section at a distance of 170 mm from point corner at different pre-stretched loading. It is seen that wrinkle amplitudes increase, but wrinkle wavelength decreases when applied load is increased. Also note that the small downwards wrinkle almost disappears, leaving an almost antisymmetric down-up wrinkle. At Fig. 9 shows the effect of pre-stretching on wrinkle formation at different location of membrane. It is observed that there is no effect on the wrinkle's amplitude at edge but all other places its increased and after greater stretching it convergence to constant.

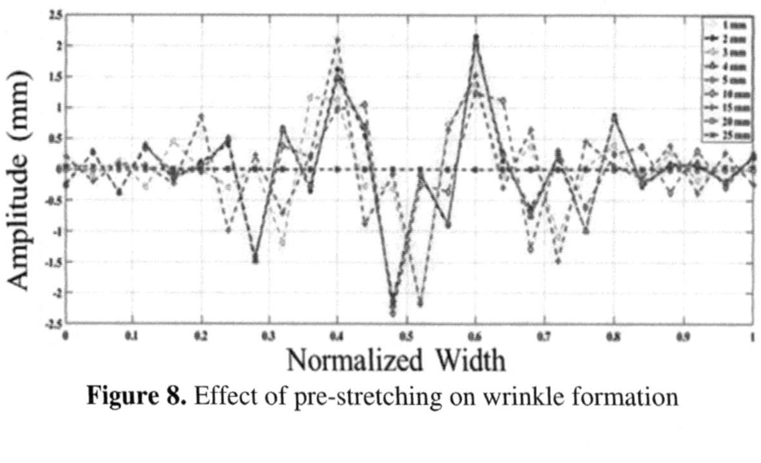

Figure 8. Effect of pre-stretching on wrinkle formation

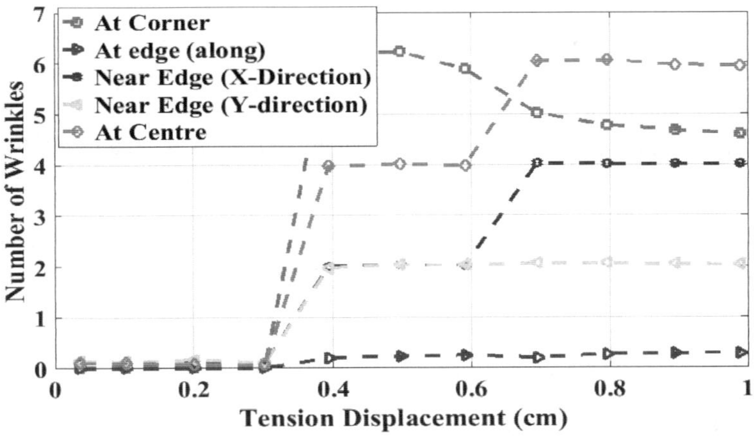

Figure 9. Effect of pre-stretching on wrinkle formation at different location of membrane

6. Conclusion

In this paper, different structural and thermomechanical condition has been discussed to analyses wrinkling behavior of membrane structures. The following conclusions have been summarized.

RMS error has been reduced by inducing the more symmetric anchor/tension points by using effective active flatness control design and double layer design. The optimum number of anchor points and respective load value can be calculated by genetic algorithms.

The scaling analysis has been performed on edge stretched membrane structures. It is found that increase in length to width aspect ratio (β) that amplitude of wrinkles also increased. The wrinkles wavelength decreases with increase in the tensile strain. But the wrinkle amplitude first increases then decreases due to increase in tensile strain. Also, it is confirmed that the wrinkle's amplitude and surface error increases with increase in length to width aspect ratio (β).

In the future, these studies will be continued in order to simulate the conditions of orbital motion.

7. Acknowledgements

The authors are truly thankful to Mr. Arup Kumar Hait, Group Head MAMG, Mr. Sammir S Sakhare, Hear AMDD, and Mr. Kripa Shankar Singh, Scientist/Engineer-SE, Space Applications Centre (ISRO) for their valuable and significant suggestions, which substantially improved the manuscript. This work supported by the Indian Space Research Organization, Government of India (Project Grant Number ISR-833-MID).

8. References

[1] C. H. M. Jenkins, "Recent Advances in Gossamer Spacecraft," *Progress in astronautics and aeronautics*, vol. 212. American Institute of Aeronautics and Astronautics, 2006.

[2] C. H. M. Jenkins, *Gossamer Spacecraft: Membrane and inflatable structures technology for space applications*, vol. 191. AIAA, 2001.

[3] E. J. Ruggiero and D. J. Inman, "Gossamer Spacecraft: Recent Trends in Design, Analysis, Experimentation, and Control," *J. Spacecr. Rockets*, vol. 43, no. 1, pp. 10–24, 2006.

[4] J. Santiago-Prowald and H. Baier, "Advances in deployable structures and surfaces for large apertures in space," *CEAS Sp. J.*, vol. 5, no. 3–4, pp. 89–115, 2013.

[5] L. Scialino, A. Ihle, M. Migliorelli, N. Gatti, L. Datashvili, K. van 't Klooster, and J. Santiago Prowald, "Large deployable reflectors for telecom and earth observation applications," *CEAS Sp. J.*, vol. 5, no. 3–4, pp. 125–146, 2013.

[6] R. E. Freeland and G. Bilyeu, "In-Step Inflatable Antenna Experiment," *Acta Astronaut.*, vol. 30, pp. 29–40, 1993.

[7] R. E. Freeland and G. Bilyeu, "In-Step Inflatable Antenna Experiment," *Int. Astronaut. Fed.*, 1992.

[8] O. Mori, H. Sawada, R. Funase, M. Morimoto, T. Endo, T. Yamamoto, Y. Tsuda, Y. Kawakatsu, J. Kawaguchi, and Y. Miyazaki, "IKAROS Demonstration Team, and Solar Sail Working Group,'First Solar Power Sail Demonstration by IKAROS.,'" *Trans. Japan Soc. Aeronaut. Sp. Sci. Aerosp. Technol. Japan*, vol. 8, 2009.

[9] A. Babuscia, T. Choi, J. Sauder, A. Chandra, and J. Thangavelautham, "Inflatable antenna for CubeSats: Development of the X-band prototype," in *IEEE Aerospace Conference Proceedings*, 2016, vol. 2016–June, pp. 1–11.

[10] N. Chahat, S. Member, R. E. Hodges, S. Member, J. Sauder, M. Thomson, E. Peral, and Y. Rahmat-samii, "CubeSat Deployable Ka-Band Mesh Reflector Antenna Development for Earth Science Missions," *IEEE Trans. Antennas Propag.*, vol. 64, no. 6, pp. 2083–2093, 2016.

[11] Y. Xu, F. Guan, H. Huang, and Q. Ye, "Thermal Distortion Analysis of Inflatable Antenna Structures Considering Inflation Gas," *J. Aerosp. Technol. Manag.*, vol. 8, no. 4, pp. 475–482, 2016.

[12] E. Cerda and L. Mahadevan, "Geometry and physics of wrinkling," *Phys. Rev. Lett.*, vol. 90, no. 7, p. 74302, 2003.

[13] E. Cerda, K. Ravi-Chandar, and L. Mahadevan, "Thin films: Wrinkling of an elastic sheet under tension," *Nature*, vol. 419, no. 6907, p. 579, 2002.

[14] M. Liu, J. Huang, and Y. Wang, "Analysis of wrinkled membrane structures based on a wrinkle-wave model," *AIP Adv.*, vol. 7, no. 1, 2017.

[15] K. Efimenko, M. Rackaitis, E. Manias, A. Vaziri, L. Mahadevan, and J. Genzer, "Nested self-similar wrinkling patterns in skins," *Nat. Mater.*, vol. 4, no. 4, p. 293, 2005.

[16] H. Soni, R. A. Pelcovits, and T. R. Powers, "Wrinkling of a thin film on a nematic liquid-crystal elastomer," *Phys. Rev. E*, vol. 94, no. 1, 2016.

[17] C. W. J. Oomens and J. D. Janssen, "The wrinkling of thin membranes: Part I—theory," *J. Appl. Mech.*, vol. 54, p. 885, 1987.

[18] D. G. Roddeman, J. Drukker, C. W. J. Oomens, and J. D. Janssen, "The wrinkling of thin membranes: Part II—numerical analysis," *J. Appl. Mech.*, vol. 54, no. 4, pp. 888–892, 1987.

[19] W. Wong and S. Pellegrino, "Wrinkled Membranes III: Numerical Simulations," *J. Mech. Mater. Struct.*, vol. 1, no. 1, pp. 63–95, 2006.

[20] Q. Tao, C. Wang, Z. Xue, Z. Xie, and H. Tan, "Wrinkling and collapse of mesh reinforced membrane inflated beam under bending," *Acta Astronaut.*, vol. 128, pp. 551–559, 2016.

[21] R. Zhao and D. Wei, "Wrinkling Analysis of Stiffness Equivalence in Membrane Structures," in *Advances in Structures, Pts 1-5*, vol. 163–167, L. J. Li, Ed. 2011, pp. 1976–1979.

[22] C. G. Wang, H. F. Tan, X. W. Du, and Z. M. Wan, "Wrinkling prediction of rectangular shell-membrane under transverse in-plane displacement," *Int. J. Solids Struct.*, vol. 44, no. 20, pp. 6507–6516, 2007.

[23] Y. W. Wong and S. Pellegrino, "WRINKLED MEMBRANES PART III : NUMERICAL SIMULATIONS," *Mech. Mater. Struct.*, vol. 1, no. 1, 2006.

[24] Y. W. Wong and S. Pellegrino, "WRINKLED MEMBRANES PART II : ANALYTICAL MODELS," *Mech. Mater. Struct.*, vol. 1, no. 1, 2006.

[25] Y. W. Wong and S. Pellegrino, "WRINKLED MEMBRANES PART I : EXPERIMENTS," *J. Mech. Mater. Struct.*, vol. 1, no. January, pp. 61–93, 2006.

[26] R. K. Miller and J. M. Hedgepeth, "An algorithm for finite element analysis of partly wrinkled membranes," *AIAA J.*, vol. 20, no. 12, pp. 1761–1763, 1982.

[27] M. Stein and J. M. Hedgepeth, "Analysis of partly wrinkled membranes," National Aeronautics and Space Administration Washington, 1961.

[28] H. Furuya, Y. Miyazaki, and Y. Akutsu, "Experiments on Static Shape Control of One-Dimensional Creased Membrane By Piezoelectric Films," in *43rd AIAA/ASME/ASCE/AHS/ASC Structures, Structural Dynamics, and Materials Conference*, 2002, no. April, pp. 1–4.

[29] N. Damil, M. Potier-Ferry, and H. Hu, "Membrane wrinkling revisited from a multi-scale point

of view," *Adv. Model. Simul. Eng. Sci.*, vol. 1, no. 1, p. 6, 2014.

[30] J. R. Blandino, J. D. Johnston, and U. K. Dharamsi, "Corner wrinkling of a square membrane due to symmetric mechanical loads," *J. Spacecr. Rockets*, vol. 39, no. 5, pp. 717–724, 2002.

[31] K. Attipou, H. Hu, F. Mohri, M. Potier-Ferry, and S. Belouettar, "Thermal wrinkling of thin membranes using a Fourier-related double scale approach," *Thin-Walled Struct.*, vol. 94, pp. 532–544, 2015.

Pavement Deterioration Model Using Markov Chain and International Roughness Index

A S Sati, S Abu Dabous, and W Zeiada

Department of Civil and Environmental Eng., University of Sharjah, Sharjah, United Arab Emirates

Email: U17105765@sharjah.ac.ae, Sabudabous@sharjah.ac.ae, wzeiada@sharjah.ac.ae

Abstract. Pavement deterioration leads to drop in serviceability and possibly failure of pavement sections due to initiation and expansion of distresses such as cracks and rutting. This paper aimed at predicting the future condition of pavement sections based on Markov chain model and the international roughness index (IRI). Developing this model can facilitate life cycle analysis and selecting the correct treatment at the right time. The historical IRI data of Canadian pavement sections were collected from Long term pavement performance (LTPP) database. IRI values were used to assess condition of the pavement sections based on the recommended ranges by the Federal highway administration (FHWA). The transition probabilities were estimated using the percentage prediction method based on historical condition data extracted from the LTPP. These probabilities are assembled in a transition probability matrix essential for the Markov chain model. The developed matrix can be used to forecast pavement conditions after any number of transition periods. The developed method assists in predicting pavement performance and facilitates the decision-making process. The method is applied to a real case study to examine its validity. The model can be expanded further by considering additional data from additional pavement networks.

1. Introduction

Pavement deterioration is the process in which the distresses develop in pavement material under the impact of environmental and traffic loading conditions. Sometimes the deterioration of pavement can lead to significant risk of increasing accidents and fatalities. Estimating performance of existing pavements and predicting their future conditions are important requirements for pavement management systems [1]. In recent years, there is a rapid growth in the pavement evaluation technology which helps a lot but at the same time complicates the process. Pavement management is the process of assessing and forecasting the pavement conditions over the service life and identifying the appropriate maintenance and rehabilitation (M&R) of roads or any paved facility. In addition, it requires a periodical collection of data from the pavement sections. Several agencies around the world have developed a road performance database. The Long-Term Pavement Performance (LTPP) which was developed in the early 1980s. It is one of the most extensive research programs supported by the Federal Highway Administration (FHWA) to collect pavement data in Canada and United State.

There are many indicators that can be used to determine the pavement condition such as pavement condition index, international roughness index (IRI), and the present serviceability index. In this research, the determination of the pavement sections condition is based on the IRI data included in the LTPP for pavement sections in Canada. The aim of this study is to develop a model that can predict the deterioration of pavement sections, the model will be developed based on the IRI data and Markov chain process. The transition probabilities are computed using the percentage prediction method.

2. Literature review

Numerous studies aimed in developing a deterioration model for many infrastructure facilities. Deterioration models are developed mainly to predict the future condition of the facility. Regarding the infrastructure deterioration models proposed Case based reasoning (CBR) to generate deterioration model of infrastructure facility with no specific facility selected [2]. Later after describing the CBR architecture, an application example was presented from the highway bridges domain. Later in the same year, the same group of researchers established a new CBR system which was developed and used in building as a proof of concept then they used it in modeling the deterioration of concrete bridge decks [3]. One year later a new approach was proposed in bridge deterioration models by identifying some categories that best represent the environmental conditions surrounded the bridge deck [4]. The proposed approach provided transportation agencies with an excellent decision support tool for bridge structures. Ranjith et al. presented Markov chain model that predict the future condition of timber bridge elements [5]. They used different methods to derive the transition probabilities and compare them to find the optimum one. Madanat et al. used Poisson regression model followed by the Markovian behaviour to calculate the transition probabilities [6]. A negative binomial regression which is a generalization of Poisson model was also used to relaxes the quality assumption between the mean and variance of the variables. Recently Markov based approach was applied to predict the future condition of water distribution network to use as a decision support system [7]. They check the validity of Markov property using Chi-square. Poisson probability distribution was used as well to compute the transition probabilities. another study was done to evaluate the fuzzy logic approach to rate the bridge conditions and to develop bridge condition rating [8]. Lastly, the fuzzy logic and probabilistic analysis was proposed to address the uncertainties in bridge condition assessment [9]. The proposed method integrated Monte Carlo simulation with evidential reasoning to enhance bridge condition rating.

Many literatures focused on pavement deterioration because of its important. Yang et al. presented an empirical study that compare the abilities of Markov chain and neural networks in modeling crack deterioration of flexible pavements using FDOT data [10]. Butt et al. developed a pavement performance and prediction model that passed on pavement condition index and the pavement age [1]. They also did a comparison between Markov model and constrained least-squares model. A series of models were developed for predicting the deterioration rate of pavement for the first 8 years of service [11]. Kobayashi et al. presented a methodology to estimate the Markov transition probabilities model to predict the deterioration of road sections [12]. Bianchini and Bandini proposed a neuro fuzzy model to predict the performance of flexible pavements using the parameters routinely collected by agencies to characterize the condition of an existing pavement [13]. Surendrakumar et al. applied the Markovian probability process to develop decision support system to expect the future condition of pavement sections [14]. Abaza (2004) developed a deterministic performance prediction model utilizes the serviceability concept adopted by AASHTO to be used in the management and rehabilitation of flexible pavements [15].

3. Data extraction

The data used in this paper was collected from the Long-Term Pavement Performance (LTPP) database which has been used by several researcher around the world. LTPP program is the largest pavement study ever conducted for US and Canada. It was initiated in 1987 as part of the Strategic Highway Research Program (SHRP) [16]. Later in 1992, The management of the program was transferred to the Federal Highway Administration (FHWA) which continue serving this role until now. The data of More than 2500 sections was collected by the LTPP program organization in cooperation with highway agencies. the data collected includes inventory, material testing, pavement performance monitoring, climate data, traffic, maintenance, and rehabilitation [16]. Before making the data available to the public, it subjected to a wide range of quality control checks. At the end, the data is stored in an information management system (the LTPP database) which is the largest pavement performance database in the world [16]. Canadian pavement sections with no maintenance and rehabilitation were collected to be used in the model development. The total number of sections was 16 sections.

From the collected data, the employed performance indicator in this study was the international roughness index (IRI). IRI describes the vehicle vibrations caused by profile roughness and is linearly proportional to roadway roughness [17]. The lower the IRI value the flatter the paved profile and vice versa. There exists no upper limit on IRI, but in practice, IRI values above 8 m/km relate to pavements nearly impassable by vehicle except at reduced speed [17]. In this paper, the mean IRI data that was collected and organized to be used in calculating the condition of each section through its service life.

4. The model

4.1. Data analysis

Before developing the model, the data should be organized to be used. Therefore, the condition changes of each pavement section were calculated using the collected IRI values. The condition will be classified into five condition states from Very good to Poor based on the IRI ranges specified by FHWA, as shown in Table 1.

Table 1. Different condition state rating based on IRI.

State	Category	IRI (m/km)	
		Interstate	Other road
1	Very good	< 0.95	< 0.95
2	Good	0.95 – 1.49	0.95 – 1.49
3	Fair	1.50 – 1.89	1.50 – 2.69
4	Mediocre	1.90 – 2.69	2.70 – 3.48
5	Poor	2.69 <	2.48 <

After calculating the condition changes of the 16 pavement sections, three sections were excluded. The reason for the first two sections was that both of them were found to be in a perfect condition during their service life where their conditions didn't drop from the very good state. The third section was excluded as the inspector started calculating the IRI data of that section after more than 10 years from its construction date, which lead to a lot of missing data. The final number of the sections that will be used to develop the model is 13 sections. The sections information is summarized in Table 2.

Table 2. Sections information.

Sec	State	Section ID	Construction date	Roadway Functional Class	Age [Yrs]
1	Ontario	A340	06/01/1981	Other	11
2	Saskatchewan	0901	09/01/1996	Other	12
3	Saskatchewan	0902	09/01/1996	Other	12
4	Saskatchewan	0903	09/01/1996	Other	12
5	Saskatchewan	0959	09/01/1996	Other	12
6	Saskatchewan	0960	09/01/1996	Other	12
7	Saskatchewan	0962	09/01/1996	Other	12
9	Quebec	A340	06/01/1981	Interstate	14
9	Ontario	2812	06/01/1981	Other	17
10	Ontario	0901	06/01/1997	Other	19
11	Ontario	0902	06/01/1997	Other	19
12	Newfoundland	1801	09/01/1984	Interstate	22
13	Quebec	3002	06/01/1979	Other	39

The historical condition state on each year for a period of 20 years of each section was summarized in Table 3. It is shown that the total number of transitions is 144.

Table 3. Historical condition states for sections.

Sec	Age [Yrs]	1	2	3	4	5	6	7	8	9	10	11	12	13	14	15	16	17	18	19	20
1	11	x	x	x	x	x	x	x	x	3	3	4	x	x	x	x	x	x	x	x	x
2	12	1	1	2	2	2	2	2	2	2	2	2	x	x	x	x	x	x	x	x	x
3	12	1	1	1	2	2	2	2	2	2	2	3	x	x	x	x	x	x	x	x	x
4	12	1	2	2	2	2	2	2	2	2	2	2	x	x	x	x	x	x	x	x	x
5	12	1	2	2	2	2	2	2	3	3	3	3	x	x	x	x	x	x	x	x	x
6	12	1	2	2	2	2	2	2	2	2	2	2	x	x	x	x	x	x	x	x	x
7	12	1	2	2	2	2	2	2	2	2	2	2	x	x	x	x	x	x	x	x	x
8	14	x	x	x	x	x	x	x	x	3	3	3	3	4	4	x	x	x	x	x	x
9	17	1	1	1	1	1	1	1	1	2	2	2	2	3	3	3	3	3	x	x	x
10	19	2	2	2	2	2	2	2	2	2	2	2	2	3	3	3	3	3	3	4	x
11	19	1	1	1	1	1	1	1	1	1	1	1	1	2	3	3	3	3	3	4	x
12	22	x	x	x	x	3	4	4	4	4	4	4	4	5	5	5	5	5	5	5	5
13	39	x	x	x	x	x	x	x	x	x	4	5	5	5	5	5	5	5	5	5	5

4.2. Markov chain framework.

There are two different basic types of deterioration models: Deterministic and stochastic. Deterministic model is basically a formula, none of the used variables are random, and the output of the model is fully determined using the parameters value. On the other hand, Stochastic model depends on random variables and represents an event where uncertainty is present. Since deterioration is an independent and random event, stochastic models are typically used. Markov chain is one of well-known stochastic model that have been used in modelling the deterioration of infrastructure facilities.

Markov chain was developed in 1906 by Andrei Markov to help decision makers analyse transition between states [7]. It is a process that has a discrete number of states, and the sequence of possible events is described. In Markov process, the probability of each event depends only on the state reached in the present regardless what was the past state as shown in Eq. 1.

$$P(X_j) = P(X_{t+1} = i_{t+1} | X_t = i_t, X_{t-1} = i_{t-1}, \dots, X_1 = i_1, X_0 = i_0) = P(X_{t+1} = i_{t+1} | X_t = i_t) \quad (1)$$

Where $P(X_j)$ is the probability of the future condition, it is the state in time t (present), and j is the future state. In Markov chain deterioration modeling the probabilities of transitioning from state i to state j called the transition probability and denoted by P_{ij}. all the transition probabilities will be presents in a matrix form. This matrix is called the transition probability matrix and denoted by P. it is the base of the Markov process. The final form of the matrix is shown in Eq. 2.

$$P = \begin{bmatrix} P_{11} & P_{12} & \cdots & P_{1j} \\ P_{21} & P_{22} & \cdots & P_{2j} \\ \vdots & \vdots & \cdots & \vdots \\ P_{i1} & P_{i2} & \cdots & P_{ij} \end{bmatrix} \quad (2)$$

The time step between states should be specified because it defines the transition period for the transition probability matrix. The transition probability matrix will represent the states and the probability of transitioning between states. The final matrix is developed for one time step. To find the transition matrix after n time step (P_t) the transition probability matrix (P) should be multiply by itself n times, or simply P can be raised to the power n as shown in Eq. 3.

$$P_t = P^n. \quad (3)$$

4.3. Percentage prediction method.

Transition probabilities of the Markov chain models can be calculated using different methods from the condition data. Some of these methods are presented in literatures such as the percentage prediction [4], regression-based optimization [1], Poisson distribution [7], and negative binomial models [6] etc. In this paper, Since the transition probabilities is computed for the sections with no maintenance and rehabilitation which is normal deterioration and developing the model will based in 13 sections only, all the sections will either transit to the next lower condition or stay in the same condition. The simple prediction method will be used to calculate the transition probabilities and develop the transition probability matrix. For other situation, other methods mentioned before can be used as well. The probability of transitioning from state i to state j can be computed using Eq. 4.

$$P_{ij} = \frac{N_{ij}}{\sum_{j=1}^{j=n} N_{ij}}. \quad (4)$$

4.4. Markov process.
After developing the transition probability matrix, it will be used to forecast the future condition through the Markov process. The present condition will be presented in terms of vector called the initial condition vector (P_0). The probability vector of transitioning to state j after n time step (P_j) will be calculated through multiplying the transition probability matrix after n time step by the initial condition vector as shown in Eq. 5.

$$P_j = P_n \times P_0. \quad (5)$$

By using P_j the expected value (Ev) of condition state can be obtained and the expected condition value of any facility after any number of years can be predicted easily by Eq. 6.

$$Ev = \sum_{j=1}^{5} j \times P_j(j) \quad . \quad (6)$$

4.5. Developing the model.
In this paper, developing the deterioration model of Canadian pavement sections using Markov process will be summarized in the following steps:

- Step 1: Select the time step, it was chosen to be one year.
- Step 2: Calculate how many transitions from state i moving to state j, summarized in Table 4.

Table 4. The total number of transitions in each state

Expected	To state				
From State	**1**	**2**	**3**	**4**	**5**
1	21	8	0	0	0
2	0	60	5	0	0
3	0	0	20	5	0
4	0	0	0	7	2
5	0	0	0	0	16

- Step 3: Calculate the transition probability using percentage prediction method, as follow:
  Pij = 21/29 = 0.72414

Make sure that the summation of each row should equal to 1. in other words, the summation of probabilities transit from state i should equal to 1.

- Step 4: Represent the transition probability in a matrix form.

$$P = \begin{bmatrix} 0.72414 & 0.27586 & 0 & 0 & 0 \\ 0 & 0.92308 & 0.07123 & 0 & 0 \\ 0 & 0 & 0.80000 & 0.20000 & 0 \\ 0 & 0 & 0 & 0.77778 & 0.22222 \\ 0 & 0 & 0 & 0 & 1 \end{bmatrix}$$

- Step 5: Use the transition probability matrix to calculate the future condition through Markov process.

After developing the transition probability matrix, the expected value of conditions will be calculated through Markov chain process mentioned before. Ten trails were done to check the validation of the model using 5 Canadian sections other than the ones used to develop the model. The information of these sections and the condition changes are summarized in Table 5.

Table 5. Historical condition for the testing sections.

Sec. ID	Age [Yrs]	1	2	3	4	5	6	7	8	9	10	11	12	13	14	15	16	17	18	19	20
1801	11					2	2	2	2	3	3	3	3	3	3	3	4	4	4	4	4
1125	12											4	4	4	5	5	5	5	5		
1645	12	1	1	1	2	2	2	2	2	2	2	2	2	2	2	2	2	2	2	2	2
A350	12						3	3	3	3	3	4	4	4	4	4	4	5	5	5	5
3016	12							5	5	5	5	5	5	5	5	5		5		5	

Before starting the calculation, it must be known that the n time step was randomly selected with taking into consideration the inspection done for how many years for each section.

- For the first trial:

- For section 1801, it was observed that in the fifth year when the inspector started the inspection, the section was initially in state 2, so this condition will be presented as a vector which is the initial condition vector $P_0 = [0\ 1\ 0\ 0\ 0]$

- Based on the section age and the inspection period, the n time step was chosen to be 10 years.

- Using the transition probability matrix computed in previous steps, the probability vector of deteriorating after 10-time steps is:

$P_j = P_{10}.\ P_0 = [0 \quad 0.4492 \quad 0.2136 \quad 0.1684 \quad 0.1688]$

- The expected value: $Ev = (1\times0) + (2\times0.4492) + (3\times0.2136) + (4\times0.1684) + (5\times0.1688) \cong 3$

- The expected condition of the section after 10 years is to be in condition 3. Table 6 contains another 9 trials that were done, and the expected values were compared with the observed values.

Table 6. Historical condition for the testing sections.

Section	n time step	Current value	Expected value	Observed value
1801	10	2	3	3
1801	15	2	4	4
1125	2	4	4	4
1125	5	4	5	5
1654	7	1	2	2
1654	15	1	3	2
A350	1	3	3	3
A350	8	3	4	4
A350	12	3	5	5
3016	3	5	5	5

From the obtained values, it was found that 90% of the results were correct. This shows that the model can be used to predict the future condition of any pavement section with a very good degree of accuracy. This can help the decision makers to predict the condition of the pavement section, rank them, and prepare any maintenance and rehabilitation needs on that time accordingly.

5. Conclusion

Predicting the future condition of pavement sections can help the decision makers to choose the appropriate maintenance or rehabilitation on the right time. This paper presented an approach for the prediction of the condition of pavement section using Markov chain and based on the IRI data extracted from the LTPP database. The data of sixteen sections in Canada with no maintenance or rehabilitation was collected. After excluding three sections due to missing information, the rest was used to develop the model. The transition probabilities were computed using the simple percentage prediction method and the transition probability matrix was developed. Five Canadian sections other than the ones used to develop the model were used to check the validation of the developed model. The results were 90% accurate compared to actual conditions reported in LTPP which showed that the developed model has sufficient degree of accuracy. To conclude, the results illustrate that Markov chain using the IRI data produced an enhanced stochastic deterioration model for pavement infrastructure. In addition, the percentage prediction method is used because of its simplicity in calculating the transition probabilities. Future work may include the expansion of the current model using additional LTPP data from USA and comparing the proposed model results with results produced by other deterioration models such as deterministic and artificial intelligence models.

6. References

[1] Butt AA, Shahin MY, Feighan KJ, Carpenter SH 1987 Pavement Performance Prediction Model Using the Markov Process Transp. Res. Rec. 12 99

[2] Morcous G, Rivard H, Hanna AM 2002 Case-Based Reasoning System for Modeling Infrastructure Deterioration *Journal of computing in civil engineering* **16** 104 114

[3] Morcous G, Rivard H, Hanna AM 2002 Modeling Bridge Deterioration Using Case-based Reasoning *Journal of Infrastructure Systems* **8** 86 95

[4] Morcous G, Lounis Z, Mirza MS 2003 Identification of Environmental Categories for Markovian Deterioration Models of Bridge Decks *Journal of Bridge Engineering* 353 361

[5] Ranjith S, Setunge S, Gravina R, Venkatesan S 2011 Deterioration Prediction of Timber Bridge Elements Using the Markov Chain *Journal of Performance of Constructed Facilities* 319 325

[6] Madanat S, Ibrahim WH 1995 Poisson regression models of infrastructure transition probabilities *Journal of Transportation Engineering* **2** 267 272

[7] Sempewo JI, Kyokaali L 2016 Prediction Of the Future Condition Of A Water Distribution Network Using A Markov Based Approach : A Case study of Kampala Water *Procedia Engineering* **154** 374 383

[8] Abu Dabous S, Alkass S 2010 A Stochastic Method for Condition Rating of Concrete Bridges *Innovation for Reshaping Construction Practice* 558 567

[9] Abu Dabous S, Al-Khayyat G 2018 A Flexible Bridge Rating Method Based on Analytical Evidential Reasoning and Monte Carlo Simulation *Advances in Civil Engineering* 2018

[10] Yang J, Lu JJ, Gunaratne M, Dietrich B 2006 Modeling Crack Deterioration of Flexible Pavements Comparison of Recurrent Markov Chains and Artificial Neural Networks *Transportation research record* 18 25

[11] Raymond C, Tighe S, Haas R, Rothenburg L 2003 Development of Canadian asphalt pavement deterioration models to benchmark performance *Canadian Journal of Civil Engineering* **30** 637 643

[12] Kobayashi K, Do M, Han D 2010 Estimation of Markovian Transition Probabilities for Pavement Deterioration Forecasting *KSCE Journal of Civil Engineering* **14** 343 351

[13] Bianchini A, Bandini P 2010 Prediction of pavement performance through neuro-fuzzy reasoning *Computer-Aided Civil and Infrastructure Engineering* **25** 39 54

[14] Surendrakumar K, Prashant N, Mayuresh P 2013 Application Of Markovian Probabilistic Process To Develop A Decision Support System For Pavement Maintenance Management *International Journal of Scientific & Technology Research* **2** 295 303

[15] Abaza KA 2004 Deterministic performance prediction model for rehabilitation and management of flexible pavement *International Journal of Pavement Engineering* **5** 111 121

[16] "LTPP InfoPave - About." [Online]. Available: https://infopave.fhwa.dot.gov/Page/Index/ABOUT_US. [Accessed: 15-Nov-2018].

[17] Park K, Thomas NE, Wayne Lee K 2007 Applicability of the International Roughness Index as a Predictor of Asphalt Pavement Condition *Journal of Transportation Engineering* **133** 706 709

IOP Conf. Series: Materials Science and Engineering **812** (2020) 012013 doi:10.1088/1757-899X/812/1/012013

Prediction of Permeability and Compressive strength for Pervious Concrete

K Pareek[1] and Y M Hong[2]

1 Student, Nanhua University, Chiayi County, Taiwan
2 Professor, Nanhua University, Chiayi County, Taiwan

E-mail: 10774107@nhu.edu.tw

Abstract. Due to characteristics of water infiltration and runoff reduction pervious concrete is becoming popular among many developed countries. The most important property of pervious concrete is its permeability and compressive strength. But it is seen that pervious concrete generally possesses low compressive strength which can be improved by various cementitious material. The aim of this research is to assess the use of silica fume, fly ash and YC819 on pervious concrete. Permeability and compressive strength test was conducted out of which compressive strength was conducted as per conventional ASTM method but permeability test was conducted by new method originated by authors. Furthermore, estimation of compressive strength using statistical regression analysis is attempted compared with model generated by different researchers. Replacement of cement by silica fume and fly ash are been done in 0 to 25% and the result showed increase in compressive strength of pervious mix by 15 to 20% whereas permeability was observed to be decreased by 3% in fly ash mixed pervious mix on the other hand silica fume reduces the permeability by 5% as compared to conventional pervious mix. In accordance with data obtained for compressive strength prediction an empirical equation was found whose R^2 was 0.67. When the models like Abrams, ACI, JSCE, Zain were applied and compared for prediction of compressive strength result showed that models do not predict the compressive strength as the different characteristics were neglected therefore it's better to use Artificial Neural Network (ANN) as compared to statistical regression model for prediction.

1. Introduction

Comprised of cement, water, aggregate and some additives in form of chemical and mineral admixtures concrete is the second most widely used substance after water on planet earth. The reason behind the concrete being the founding material for the modern development lies in the fact of its abundance in nature (limestone), durability, thermal mass and ability to retain storm water etc. There are several factors on which quality of concrete is determined such as water cement ratio, coarse aggregate size, fine aggregate size, curing condition and etc. [1,2].

Concrete is used in construction because of its compressive strength which is very high as compared to its tensile strength. Generally determined by cubes or cylinder under 28 days curing. In case of cylindrical specimen 150 mm diameter and 300 mm height a high load is applied longitudinally under uniaxial compression loading which is then divided by loading area to get compression capacity [3]. On the other hand, permeability is defined as the movement of fluid through a porous medium under an applied pressure head is the most important property of concrete governing its long-term durability. Permeability of concrete, in turn, is influenced by two primary factors: porosity and interconnectivity of pores in the cement paste and micro-cracks in the concrete, especially at the paste–aggregate interface. Porosity and interconnectivity are controlled for most part by the w/c ratio, degree of hydration, and the degree of compaction [4]. In any type of concrete construction and engineering judgement prediction of compressive strength and permeability is becoming a necessary requirement. In last couple of decade compressive strength prediction has reached new stage with application of SPSS and ANN. Statistical Package for the Social Sciences shortly known as SPSS is a user friendly yet powerful software package used for analysing, manipulating and presenting data which is mostly used in behavioural and social sciences. Among the different forms of SPSS programs available the core program is known as SPSS Base which is further extended into multiple modules on the basis of varying statistical and data entry capabilities [5].

In this study Regression Model SPSS is used as it helps to form a relationship among the variable input. Moreover, it helps to understand the variable change in dependent value keeping some independent value fix and other changeable. But in most of the cases, the estimation target is a function of the independent variable known as the regression function. The aim of this study is to determine the compressive strength and permeability of pervious concrete mix and then to analysis the effect of different parameters like water, cement, coarse aggregate etc. on compressive strength by using multiple linear regression. At last an empirical equation is obtained based on the analysis result which is used to compare with actual results. The empirical formula is then used to compare the results of other researchers stating its applicability.

2. Experimental work

The specific and detailed sets of data should be provided to SPSS to predict the permeability and compressive strength of pervious concrete mix. In this study 915 concrete specimens were used out of which some concrete specimens are prepared at Nanhua University laboratory and some of data set are obtained from Professor ChenYeh from Chung-Hua University [6]. Reason for adding more data correspond to fact that geographic position of both the places are nearly same as well as extraction of materials are quite similar. The specimens were constructed in cylindrical shape with a diameter 15 cm and height of 30 cm. Table 1 depicts the characteristics of concrete data used in the experiment and table 2 depicts the number of patterns used for SPSS modelling. Details of basic materials used is shown below:

- Cement- Ordinary Portland type-1 which is available in local Taiwanese market is used. The bulk density of cement type used is 1257 kg/m^3.
- Coarse Aggregate- 10 mm and 20 mm are the two type of CA were used in this study which are available in Taiwanese local market. The value of bulk density of CA used is 1623 kg/m^3.
- Water- Potable water was used for casting of all the specimens
- Fly Ash- Fly ash was obtained from Guitian Lab Doliu, Taiwan. Fly ash was used to replace various percentage of weight proportions of cement. The specific gravity of fly ash is 2.7.
- Silica fume- It was procured from Guitian lab, Taiwan. Silica fume was also used to replace various percentage of weight percentage of cement. The specific gravity of silica fume is 2.2

Table 1. Characteristics of concrete data

Serial Number	Parameters	Minimum	Maximum	Unit
1	Age	7	365	days
2	Blast Furnance Slag	0	359.4	kg/m^3
3	Cement	102	540	kg/m^3
4	Coarse Aggregate	801	1299.1	kg/m^3
5	Compressive strength	7.51	82.6	MPa
6	Fine Aggregate	594	992.6	kg/m^3
7	Fly Ash	0	483	kg/m^3
8	Silica Fume	0	483	kg/m^3
9	Super Plasticizer	0	32.5	kg/m^3
10	Water	72.2	247	kg/m^3

Absolute volume mix method was used to find out the concrete mix proportions. It is accurate method for making concrete specimen occupying unit volume using specific gravity for all the ingredients used. Compression strength test was conducted on the basis of ASTM C39 at the age of 28 days of curing and was measured through Universal testing machine. Table 3 depicts the characteristics of specimens constructed and used for regression analysis.

Table 2. Number of patterns used in Regression Modelling

Serial number	Number of patterns	Compressive Strength (MPa)
1	112	7 to 20
2	185	20 to 30
3	239	30 to 40
4	164	40 to 50
5	116	50 to 60
6	58	60 to 70
7	36	70 to 83

Specimens prepared at Nanhua University laboratory were subjected to Permeability measurement using storm water pipe having diameter of 15.1 cm and having 40 cm height as shown in figure 1 as a sample placement. In this test we pour 1litre of water through sample and after 1 minute, we how water is retained on the base below. Same procedure is reiterated for 6 times until the reading get constant which indicate that all the unfilled air voids are full and the water flows only through voids interconnected voids of concrete. Sample picture of the setup made for calculation of permeability has been shown in figure 2. All test was done in accordance with IS 516 and table 2 below shows the value of average permeability recorded.

Figure 1: Permeability test setup

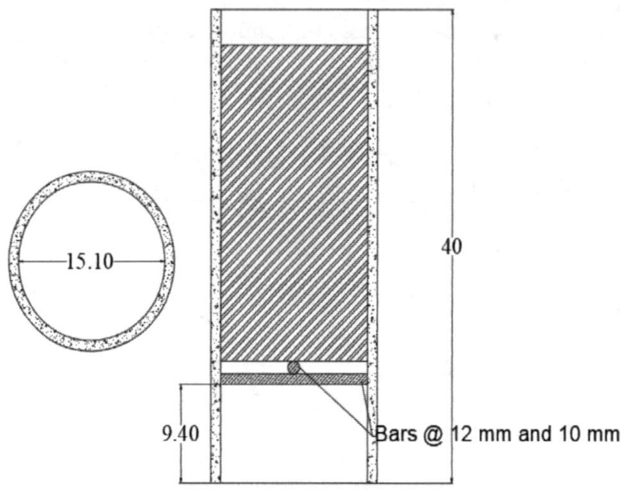

Figure 2: Cross sectional and side view of experimental setup (dimensions are in cm

Table 3: Depicting the characteristics of specimens

Aggregate size (mm)	Water cement ratio	Cement (kg/m^3)	Aggregate quantity (kg/m^3)	Water (kg/m^3)	Admixture type	Admixture quantity	Sample Code
20	0.35	250.9	1299.09	117	Nil		M1
20	0.40	250.9	1299.09	100.75	Nil		M2
10	0.35	250.9	1272.18	117	Nil		M3
10	0.40	250.9	1272.18	100.75	Nil		M4
20	0.35	225.81	1299.09	117	Fly ash	25.09	M5
20	0.40	225.81	1299.09	100.75	Fly ash	25.09	M6
10	0.35	225.81	1272.18	117	Fly ash	25.09	M7
10	0.40	225.81	1272.18	100.75	Fly ash	25.09	M8
20	0.35	250.9	1299.09	84.5	YC819	32.5	M9
10	0.40	250.9	1272.18	87.75	YC819	13	M10
10	0.35	250.9	1272.18	104	YC819	13	M11
20	0.40	250.9	1299.09	87.75	YC819	13	M12
20	0.35	250.9	1299.09	104	YC819	13	M13
10	0.40	225.81	1272.18	100.75	Silica fume	25.09	M14
10	0.35	225.81	1272.18	117	Silica fume	25.09	M15

IOP Conf. Series: Materials Science and Engineering **812** (2020) 012013 doi:10.1088/1757-899X/812/1/012013

20	0.40	225.81	1299.09	100.75	Silica fume	25.09	M16
20	0.35	225.81	1299.09	117	Silica fume	25.09	M17
10	0.40	250.9	1272.18	72.15	YC819	28.6	M18
10	0.40	188.11	1272.18	100.75	Fly ash	62.79	M19
10	0.35	188.11	1272.18	117	Fly ash	62.79	M20
10	0.40	188.11	1272.18	100.75	Silica fume	62.79	M21

3. Regression analysis and performance criteria

Statistical Regression Analysis is a statistical tool used to predict the relationship between a dependent variable and an independent variable. This relationship is generally expressed through statistical model equation through which we can predict the future value of response variable using regression variable, note that we generally use statistical method when there is continuous dependent variable and number of independent variables. But these parameters are estimated so that a measure of fit is optimized having values as close as possible to real system [7,8]. There are generally two type of regression analysis model linear regression model and multiple regression model. Linear regression is most common form of regression and most common type of linear regression is least square regression. In this we basically use two sets of given values to predict the futuristic value of dependent. R^2 is measure of association which represent the variance in value of dependent variable that can explained by knowing the value of independent variable. Another type of regression analysis model is multiple regression model which is an extension of linear regression. Multiple regression allows additional factors to enter the analysis separately such that effect of each is analysed [5,9]. The aim of using multiple regression is to enable the researcher to assess the relationship between more than three independent variables with dependent variable. Value of R^2 generally varies from 0 to 1.

Statistical regression analysis is a powerful tool to predict the compressive strength of concrete considering the materials input. The origin of predicting compressive strength starts with Abrams law which stated that increasing the water cement ratio is inversely proportional to compressive strength [10]. Abrams law served as base research for many researches such as Zain and Zhang. American Concrete Institute also gave an equation stating the relation between compressive strength and water cement ratio [11]. But both Abrams law and ACI equation didn't mention about time of curing, use of chemical and natural admixture. Zain et al also used multiple regression model to predict the compressive strength of high performance concrete and yielded an excellent correlation coefficient of 99.99% accuracy of concrete specimen using mineral admixture like silica fume (SF) and fly ash (FA) at the different ages of curing (3,7,14, 28 and 81) [12]. In 2011 Zhang et al. stated that prediction of compressive strength by varying the time period from 7 to 365 days [13]. JSCE (2012) stated that prediction of the compressive strength using the minerals admixtures and time period at the same time but in specified series of dates [14]. In table 4 comparative analysis has been done depicting the empirical equations given by authors and institutes whereas table 5 and table 6 represent the correlation coefficient founded by authors and institutes.

In this study an attempt is made to predict the compressive strength by varying the mineral admixtures concentration with time as a function of compressive strength. Multiple linear regression is used to predict the compressive strength of cylindrical specimens. The result of the analysis is the empirical formula between the compressive strength and various parameter. The various parameters are the cement concentration, water concentration, silica fume, blast furnance slurry, fly ash, coarse aggregate, fine aggregate and w/c ratio. By using SPSS software version an empirical formula between the compressive strength of concrete (f_{cm}) as dependent variable and cement (C), water (W), fine aggregate (fine agg.), coarse aggregate (Coarse agg.), blast furnance slag (BFS), fly ash (FA), super plasticizer (SP) and age (curing age) as an independent variable. Analysis results also known as the R^2 value. R^2 is always between 0 and 100%.

Table 4: Equations to predict the compressive strength of concrete cylinder where f_{cm} is average compressive strength of the concrete cylinder (150mmx300mm)

Author	Formula proposed	Coefficient values and limitations
Abram	$fcm = \frac{A}{B^\wedge w/c}$	The constants A and B are 96 MPa and 7, respectively. w/c ratio should be between 0.3 and 1.2
American Concrete Institute Manual of Concrete Practice (ACI2000-I)	$fcm = 117.07 \cdot e\text{-}2.572 \cdot w/c$	The w/c ratio and cement content of the concrete mixes are limited to 0.41–0.82 and 300–360 kg/m3, respectively. The equation was adapted from the results of the table given in the specification with R2 = 0.996
Zhang (2011)	$fcm = \left(\frac{22.5C}{W} - 12.1\right)[(1.05r + 0.15)lnt - 2.77r + 0.79)]$	r=fly ash upto 50% replacement and t indicate time (7<t≤365)
Zain (2009)	$f_{cm}= a_0C^{a}{}_{1}.W^{a}{}_{2}.Fa^{a}{}_{5}.Ca^{a}{}_{6}.SF^{a}{}_{7}.\rho^{a}{}_{8}.FA^{a}{}_{9}.S^{a}{}_{10}$	The value of the different constant varies according to number of day
Japanese Society of Civil Engineering (JSCE)	$fcm = a\left(\frac{C}{W} + K.\frac{F}{W}\right) + b$	Constant a, K, b whose values are shown in table, where replacement limit for fly ash is 40%

Table 5: Correlation coefficient between compressive strength at different stage and selected variables for model proposed by Zain

Variable	3 days CS	7 days CS	14 days CS	28 days CS	81 days CS
Ca	-0.99	-0.99	-0.99	-0.99	-0.99
Fa	-0.98	-0.98	-0.99	-0.99	-0.98
C	0.92	0.89	0.91	0.91	0.85
SF	0.22	0.28	0.22	0.2	0.36
FA	0.1	0.09	0.12	0.12	0.08
W	-0.99	-0.98	-0.99	-1	0.97
S	0.59	0.57	0.56	0.56	0.58
ρ	0.26	0.2	0.24	0.25	0.13

Table 6: JSCE equation correlation coefficient value

Age (days)	K	a	b
7	0.253	24.016	-16.756
28	0.390	28.938	-15.616
91	0.549	32.243	-13.939

4. Results and discussion

We check the permeability of plain pervious concrete cube 1000 ml water which is passed through the voids of the plain pervious concrete cube and the water is retained to another pan that stored it and then measured. Eight times the reading were taken with the gap of 3 min and average of which is shown in table below. The universal testing machine is used for testing the compressive strength of concrete according to ASTM C39M/2003 for cylinders (f_{cm}) measuring 150mm x 300mm. Testing was carried out at 28 days and at age of testing.

Result obtained from the above experiments shows that replacement of Silica fume and fly ash have negative effect on permeability but have affirmative effect on the compressive strength of mix. With increase in replacement of fly ash from 0% to 25% there was 3% decrement seen in the permeability of the mix on the other hand compressive strength of the mix increases by 15%. When Silica fume is used as admixture compressive strength of mix increases upto replacement of 25% but further more increase in silica fume concentration reduces the permeability by 5% as compared to conventional mix. While permeability reading doesn't varied that much. When YC819 is used as the admixture the setting time of concrete mix increases but the permeability of mix remains but compressive strength of mix was observed to decreased with increase in concentration of YC819. Aggregate size have also deep effect as with 20mm aggregate permeability is more compared to 10 mm based concrete mix can be observed from figure 3. Based on above results, we found that use of silica fume should be confined to 5% when mixed in pervious concrete as it enhances the compressive strength but decreases the permeability of the mix. For fly ash replacement should be done upto 10% as it increases compressive strength and permeability also remains constant compared to 0% mix.

The data obtained from compressive strength test and from prof Yeh are used for mathematical modelling by using multiple linear regression method. The purpose of mathematical modelling was to get an empirical formula to establish the relationship between compressive strength and variables like cement, water, coarse aggregate, fine aggregate, fly ash, super plasticizer, age of curing and blast furnance slag. The coefficient of determination (R^2) is the proportion of the total variation in f_{cm} explained by the regression of f_{cm} on (X_1-X_8) and R^2 ranges from 0 to 1 i.e. when the estimated regression allows none of variation in f_{cm} to the stage when all points lie on the regression line. Analysis was done by using SPSS program version 22. The empirical formula predicting the compressive strength is as below:

$$f_{cm} = C + A_1X_1 + A_2X_2 + A_3X_3 + A_4X_4 + A_5X_5 + A_6X_6 + A_7X_7 + A_8X_8$$

Where f_{cm} is the compressive strength of concrete in N/mm^2, X_1 is cement in kg/m^3, X_2 is blast furnance slag in kg/m^3, X_3 is fly ash in kg/m^3, X_4 is water kg/m^3, X_5 is super plasticizer in kg/m^3, X_6 is coarse aggregate in kg/m^3, X_7 is fine aggregate in kg/m^3, X_8 is age in days respectively.

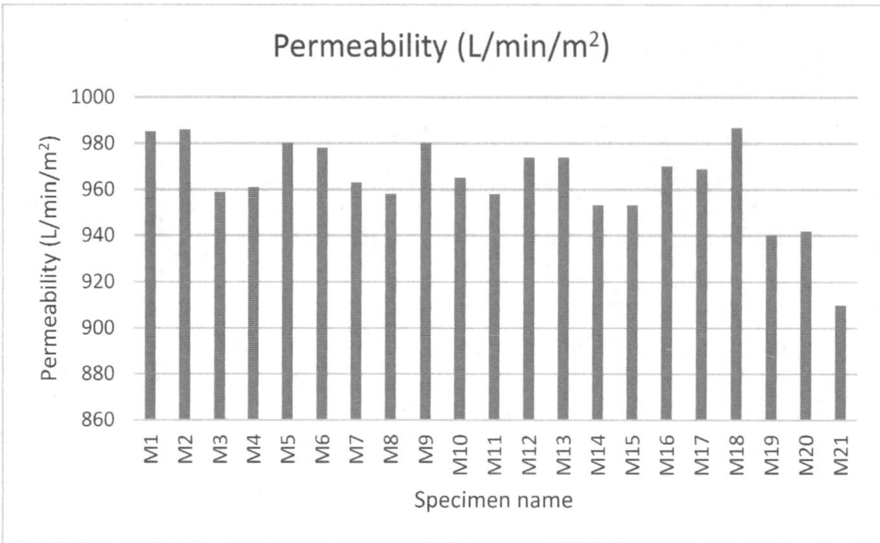

Figure 3: Permeability value comparison

Through obtained empirical equation for test data comparison of estimated values with experimental compressive strength value have been given in figure 4. Complete data set were then used to testify the equation suggested by different researchers. The original method of Abrams, ACI and modified method of JSCE, Zhang & Zain cannot accurately predict the compressive strength of cylinder specimen as the original method uses only water-cement ratio as an influencing factor and on the other hand modified method consider different factors but didn't uses the all the factor at the same time. In other words, the estimated strength of concrete based on the quality of cement paste is not accurate, and the relationship between the actual and calculated strength for all the original models is poor when only w/c (with and without considering the volume of aggregates to the cement content) is considered as an influencing factor. Even if empirical equation given have better R^2 of 0.80 or even 0.90 it is not the same for other researcher data. Figure 5 compares the empirical formula given by researchers and it can be observed from the figure that these empirical equations cannot accurately predict the compressive strength accurately due to reasons mentioned above. Therefore, it is suggested to use different computing approach such as ANN for predicting the compressive strength.

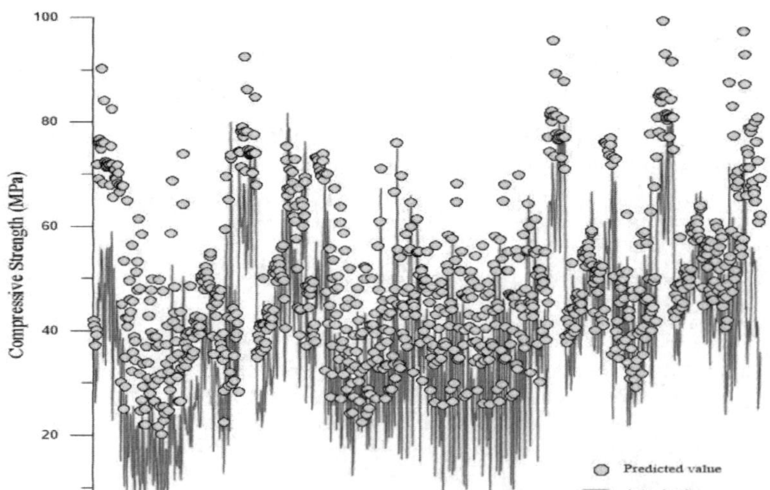

Figure 4: Comparison of result between experimental value and predicted value

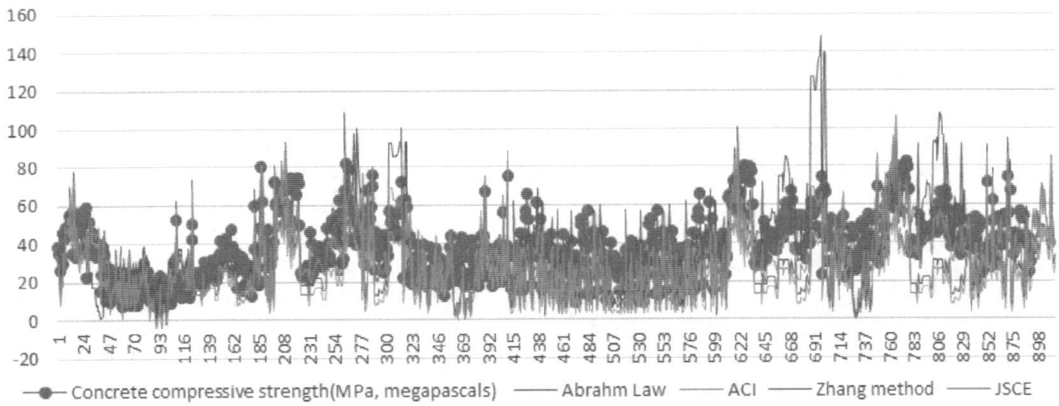

Figure 5: Comparison of actual and calculated compressive strength of concrete mixes with the equation as suggested by different authors

5. Conclusion

For achieving high permeability and porosity pervious concrete are favoured as compared to any other concrete type. Pervious concrete has 15 to 30% more pores as compared to conventional concrete but that lowers its compressive strength many folds. Therefore, conventional pervious concrete mix requires some modifications like addition of admixtures like fly ash, silica fume and YC819. When fly ash, silica fume and YC819 were used following conclusions can be made considering permeability as main factor:

- Replacement of fly ash should be done upto 15% as it increases the compressive strength and doesn't alter the permeability of the mix compared to conventional pervious mix.
- Application of silica fume should be limited to 5% as it increases the compressive strength but decreases the permeability.
- YC819 have nearly no observed effect on permeability on pervious mix but decreases the compressive strength of mix.

Empirical equation obtained from regression analysis shows great potential but not at required level as R^2 of the equation was 0.67 only which shows huge gap between predicted and actual value of compressive strength. When the empirical equation of different researchers was used huge variation of compressive strength were recorded. Therefore, it is suggested to use artificial neural network (ANN) for predicting the compressive strength.

6. References

[1] S SP 1993 Recent trends in the science and technology of concrete, concrete technology, new trends, industrial applications *Proc. of the international RILEM workshop, London, E & FN Spon*, pp. 1–18.

[2] B HH 1981 Densified cement/ultrafine particle based materials *The second international conference on superplasticizers in concrete*, Ottawa

[3] Bhanja S and Sengupta B 2002 Investigations on the compressive strength of silica fume concrete using statistical methods *Cement and Concrete Research*, vol. 32, pp. 1391-1394

[4] Banthia N, Biparva A and Mindess S 2005 Permeability of concrete under stress *Cement and Concrete Research* vol. 35, pp 1651– 1655

[5] Pitroda J Prediction of strength for fly ash cement concrete through soft computing approaches *International Journal of Advance research in engineering, Science and management (IJARESM)*, vol. 1, pp 1-11

[6] Yeh I.-C. 1998 Modeling of strength of high performance concrete using artificial neural networks *Cement Concrete Research* vol. 28(12), pp 1797–1808

[7] Chopra P Sharma R K and Kumar M 2014 Regression Models for the Prediction of Compressive Strength of Concrete with & without Fly ash *International Journal of Latest Trends in Engineering and Technology (IJLTET)*, Vol. 3 Issue 4, pp 400-406

[8] Maruthachalam D, Rajalaxmi R K and Vishnuram B G 2012 Statistical Modelling of Fiber Reinfored High Performance Concrete *International Journal of Scientific & Engineering Research*, Vol 3 No.6, pp.1-5

[9] Deshpande N, Londhe S and Kulkarni S S 2013 Modelling compressive strength of recycled aggregate concrete using neural networks and regression, *Concrete Research Letters*, Vol. 4 No.2, pp.580-590.

[10] Abrams, 1919 L.D. Properties of Concrete, 3rd ed.; Pitman Publishing Ltd.: London, UK

[11] ACI 2000-I. ACI Manual of Concrete Practice 2000, Part 1: Materials and General Properties of Concrete; American Concrete Institute (ACI): Farmington Hills, MI, USA

[12] Zhang W, Yoshitake I and Saitoh T 2011 Proposal of a Simplified Prediction Formula for Compressive Strength of Fly Ash Concrete *Advanced Materials Research* Vols. 287-290 pp 1201-1208

[13] Zain M F D and Abd S M 2009 Multiple regression model for compressive strength prediction of high performance concrete *Journal of Applied Sciences* Vol. 9(1) pp 155-160.

[14] Japan Society of Civil Engineers (JSCE) 2009: New Utilization Technologies of Fly Ash Concrete Suitable for Recycle Oriented Society-Recommendations for Design and Construction of Fly Ash Concrete (Draft), Concrete Library 132, pp.83-87.

IOP Conf. Series: Materials Science and Engineering **812** (2020) 012014 doi:10.1088/1757-899X/812/1/012014

A wrinkling analysis and control of rectangular membrane structures with upscale modelling under on-orbit conditions

S Kumar[1,a], S H Upadhyay [2,b], V V Koryanov[3,c]

[1]BIT Sindri, Dhanbad-828123, India
[2]Smart Materials and Structures Laboratory, Department of Mechanical & Industrial Engineering Indian Institute of Technology, Roorkee-247667, India
[3]Dynamics and flight control of rockets and spacecraft department Bauman Moscow State Technical University, Moscow, Russia, 105005

Email: [a] skumarnitj@gmail.com, [b] shumefme@iitr.ac.in, [c] vkoryanov@bmstu.ru

Abstract. This paper presents detailed analysis of a novel concept on the formation and evolution of wrinkle pattern on a multilayer shell-membrane under the action of mechanical and thermal-mechanical loading. The double scale Fourier series is used to deduce the wrinkling equation of membrane structure. In this paper discusses the theoretical aspects of this process and carried out part of the numerical calculations. In particular: wrinkling analysis of square membrane with corner tensile loading. It contains: schematic loading condition on square membrane, membrane with shell elements and experimental analysis of square membrane with tensile loading. Also held stress analysis. In the next paper, the authors will consider a detailed analysis of the research.

1. Introduction

The large space structures are required for the deep space exploration, high speed communication, and accurate earth observation. But the weight of the payload increases when the size of structures increase in the conventional space structures and probability of the successes decreases. Also, the operation cost required to transport mass to the destination is high as well as the volume of payload is limited. Hence, achieving minimal launch mass and volume of space structures is always an important task for the space mission. [1]–[5]. Therefore, ultra-lightweight and ultra-large membrane structures are the key element of inflatable reflector, solar sail, solar collector, sun-shields, deployable mirrors surface for the next generation of spacecraft. Many such membrane based space structures have been tested like IN-Step Inflatable antenna experiment [6], [7], IKAROS [8], CubeSat [9], [10].

Thin membranes are often subjected to stresses and strains when an external force applied. Membrane structure is very flexible and has strong geometric nonlinearity and can leads to an out of plane displacement due to their negligible flexural rigidity and presence of in-plane compressive stresses [2], [11]–[14]. Hence it is essential and desirable for a better understanding of the effects of wrinkles on the structural performance and its stability. Surface wrinkling may also be utilized to develop techniques for probing the surface characteristics of materials. [13], [15], [16].

Many authors have discussed the various analytical, numerical and experimental methods to study the wrinkling behavior of thin membrane structures [16]–[25]. Reissner (1938) derived the tension field theory without initial tension under in-plane torsion to and analyzed the circular membrane [26]. Stein and Hedgepeth (1961) to analyze partly wrinkled membranes using Stein's theory [27]. Using the Rayleigh-Ritz method, the circular stretched membrane is analyzed without bending rigidity, but geometrical stiffness considered by [28]. Damil et al [29] have discussed a wrinkling phenomenon using a new multiscale approach taking one-dimensional beam model. Blandino et al [30] have discussed the wrinkling phenomena at the isothermal condition and prove that number of wrinkles increase with the tensile loads. Also, the amplitude of wrinkles increases with increase in thermal gradient. The full shell methods and the reduced resolution methods can be used to solve the wrinkle problems. For the thermo-mechanical load, Kodjo et al [31] have modified Roddeman's model [18] to considering thermal effect in the model.

In this paper, thermal effect on flat membrane antenna has been discussed at orbital conditions. Also, authors discussed the wrinkling analysis of the membrane structures under different loading conditions, the effect of prestress, thickness variation, elastic properties, anchor points. Thermal stresses have been introduced in the classical equation of within the framework of the elastic shell theory. Here Asymptotic Numerical Method is used to solve the nonlinear equations and effective active flatness control method is used to minimize the RMS error of the surface wrinkles of the membrane structure. Also, the analysis of bi-layer structure has been performed to minimize the wrinkled area.

2. The mathematical formulation of nonlinear model of coupled thermal and elastic behavior

Current research discussion is focused on to develop a nonlinear model of thin membrane structures under mechanical and thermal loads. The Love–Kirchhoff plate theory is used to derive the equations of the wrinkling model within the von Karman approximation. The membrane model is considered as the elastic and isotropic. The thermal stress has been solving by applying the generalized Duhamel-Neumann form of the Hooke's law. For the single element thermal strain, ε_{therm} under temperature change is defined as,

$$\varepsilon_{therm} = \alpha\,\Delta T \tag{1}$$

The relations of linear strain-displacement (Kinematical) are well known as,

$$\begin{aligned} \varepsilon_{xx} &= u_{,x} \\ \varepsilon_{yy} &= v_{,y} \\ \gamma_{xy} &= u_{,y} + v_{,x} \end{aligned} \tag{2}$$

Where, $(\cdot)_{,x} \equiv d(\cdot)/dx$ and $(\cdot)_{,y} \equiv d(\cdot)/dy$.

Let the displacement field be represented by

$$\begin{aligned} u &= \wp_i u_i \\ v &= \wp_i v_i \end{aligned} \tag{3}$$

Where 'i' is the subscript i (i=1,2,3,..) applies for summation convention and u(x, y), v(x, y) and o(x, y) stand for the displacement functions. $\wp_i(x, y)$ is the longitudinal variation in u(x, y), v(x, y) and w (x, y) along x and y-direction. From equation (2) and (3) we can obtain the strain components as

$$\begin{aligned} \varepsilon_{xx} &= u_i \wp_{i,x} \\ \varepsilon_{yy} &= v_{i,y} \wp_i \\ \gamma_{xy} &= u_{i,y} \wp_i + v_i \wp_{i,x} \end{aligned} \tag{4}$$

Now the constitutive (stress-strain) relations of the isotropic sheet, under plane stress can be written as,

$$\begin{aligned} \sigma_{xx} &= \Im_{xx}\varepsilon_{xx} + \Im_{xy}\varepsilon_{yy} \\ \sigma_{yy} &= \Im_{yx}\varepsilon_{xx} + \Im_{yy}\varepsilon_{yy} \\ \sigma_{xy} &= G\gamma_{xy} \end{aligned} \tag{5}$$

$$\Im_{xx} = \Im_{yy} = \frac{E}{1-\nu^2}, \quad \Im_{xy} = \Im_{yx} = \frac{\nu E}{1-\nu^2} \tag{6}$$

Where E is the Young's Modulus, ν is the Poisson's ratio and $G = E/2(1+\nu)$ is the shear modulus. On the basis of Fourier-related approach [32] the local unknown filed can be written as

$$\mathbf{U}(x, y) = \sum_{l=-\infty}^{+\infty} \mathbf{U}_l(x, y)\exp(ilRx) \tag{7}$$

Where R is the wave number and $\mathbf{U}_l(x, y)$ the unknown field slowly varies on the period $\left[x, y+\frac{2\pi}{R}\right]$ of the oscillation. Using equation (7) the useful formula of membrane strain equation [29], [32] can be derived in as

$$\{\gamma\} = \begin{Bmatrix} \frac{\partial u}{\partial x} + \left(\frac{\partial w}{\partial x}\right)^2 + R^2 w^2 \\ \frac{\partial v}{\partial y} + \left(\frac{\partial w}{\partial y}\right)^2 \\ \frac{\partial u}{\partial y} + \frac{\partial v}{\partial x} + 2\frac{\partial w}{\partial x}\frac{\partial w}{\partial y} \end{Bmatrix} \tag{8}$$

The membrane strain formula Eq. (8) is similar to the Von Karman model. It contains two-part symmetric part of membrane strain and second part of the nonlinear part of wrinkling deformation and is given as

$$\{\gamma^{wr}(w)\} = \begin{Bmatrix} \left(\frac{\partial w}{\partial x}\right)^2 + R^2 w^2 \\ \left(\frac{\partial w}{\partial y}\right)^2 \\ 2\frac{\partial w}{\partial x}\frac{\partial w}{\partial y} \end{Bmatrix} \tag{9}$$

The wrinkling initiation in the bilayer structures can describe with the equations of the critical strain, wavelength, and wave number is sufficient and quite simple. The solution of bilayer structures should be extended to multilayer by following a simple intuitive procedure. The equation of wave number by relating the axial strain and to wave number for the bilayer wrinkling has been derived similarly to [33] as,

$$\varepsilon_f \frac{t_f^3}{12}\ell^4 - \varepsilon_f t_f \gamma \ell^2 = -\frac{2\varepsilon_s}{\xi}\ell \tag{10}$$

Where ℓ is the wave number, ε_f and ε_s refer to the plane strain of the film of upper layer lower layer. ν_s is the Poison ratio of the substrate, t_f is the film thickness, and $\xi = (3-4\nu_s)/(1-\nu_s)^2$.

Material Properties. Many materials with improved combinations of properties are used by space scientist/engineer to select their specific mission requirement [2, 5, 7, 9]. Membrane structures consist of two-dimensional pre-stress, elastic thin membrane as a major structural element. The membrane mentioned in this paper is different from the physical "true membrane". Basically, bending and compressive rigidity does not found in the true membrane. Also, structures made from true membrane does not have structure stiffness. Shell-membrane has stiffness comes from initial pre-stress, not from the material and shape [20], [22]. In practice, any two-dimensional elastic continuum resists bending moment. However, if the tension is large and the curvatures are small, the effect of bending moment can be neglected. Thus, the membrane can be imagined as an extension of the string to two dimensions [19].

For the analysis of membrane structure following assumptions are made as; i) Effect of gravity on the membrane is negligible. ii) Displacement is only in vertical direction. iii) The membrane is thin enough to neglect its volume and only consider its area. iv) The magnitude of pre-stress remains constant and mass density assumed uniform throughout the membrane.

It is perceived that Kapton is one of the best-qualified materials for the space application, which has better mechanical and thermal properties like strength is to weight ratio, tensile strength, etc. The Material properties of Kapton®HN is as Density is 1420 kg/m^3, Young's Modulus is 2.50 x 10^9 N/m^2, Poisson's Ratio is 0.34, Tensile Strength is 231 x 10^6 N/m^2, Coefficient of Thermal Expansion (α) is 20 ppm, Specific Heat is 1090 J/kg-K, Thermal Conductivity is 0.12 W/m K.

3. Numerical Examples

In this section several examples are presented for the nonlinear buckling element model is performed to wrinkling analysis of a thin membrane under mechanical and thermal loads. Characterization of the wrinkling behavior of the membrane structures has been carried out in these examples. The nonlinear buckling finite element analysis incorporating thin shell model is used to simulate the detailed wrinkles in the membrane structure. Elastic quasi-static shell analysis and parameter studies of the thin-film membrane are carried out using the geometrically nonlinear updated Lagrangian description and a displacement-based isoparametric finite element formulation. The nonlinear equation is solved by using the Newton-Raphson iteration and dichotomy method as [22], [34]. The macroscopic equation of reduced membrane model is implemented in Abaqus subroutine using Fortran compiler. In their simplest form, these schemes begin by assuming that the behavior of the membrane is linear elastic. Then, any compressive principal stresses are set to zero and the associated stiffness matrix coefficients are also set to zero. The principal stresses are recalculated at every iteration, to eschew history dependency in the results.

4. Wrinkling Analysis of Square Membrane with Corner Tensile Loading

In this section, wrinkling analysis of membrane under corner tensile loading is discussed as Fig. 1 (a). All corners are biaxial loaded by Kevlar tread. Kapton tape is used to hold the membrane under tension condition with Kevlar tread. Threads are directly attached to force sensors to measure the applied tension load as Fig. 2. The membrane and corner tabs are modeled using S4R thin shell elements of different thickness and corner beams are modeled using B21 beam elements with the general beam section. Mesh convergence shows that 2500 S4R shell elements are required for wrinkling analysis of square membrane (as Fig. 1 (b)). The *MPC, TIE function is used to connect each beam node to the corresponding shell element node. The membrane is constrained in both x and y-directions at the center node; all four side edges are left free; the z-translations and all rotational degrees of freedom of the corner beam nodes are restrained. The corner loads are distributed uniformly over the nodes of the beams (with only half of the forces applied at the end nodes). The analysis procedure is essentially identical for all of the simulations. First, a uniform pre-stretch of 0.1 mm is applied to provide some initial out-of-plane stiffness to the membrane. This is achieved by using the *INITIAL CONDITION, TYPE=STRESS parameter in ABAQUS. A nonlinear geometry analysis is then carried out, with the *NLGEOM option activated, to check the equilibrium of the system with this initial pre-stress. The chosen geometrical imperfections are then seeded onto the pristine mesh using the *IMPERFECTION command. The out-of-plane displacements following loading are obtained, showing the number of wrinkles, their amplitudes and their locations on the membrane.

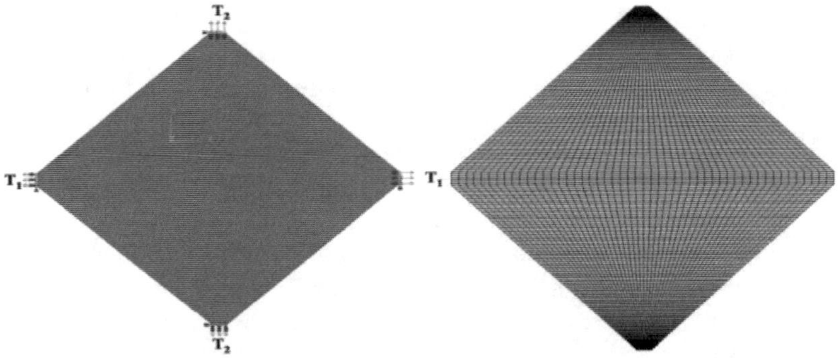

(a) (b)

Figure 1. a) Schematic loading condition on square membrane,
b) Membrane with 2500 S4R shell elements

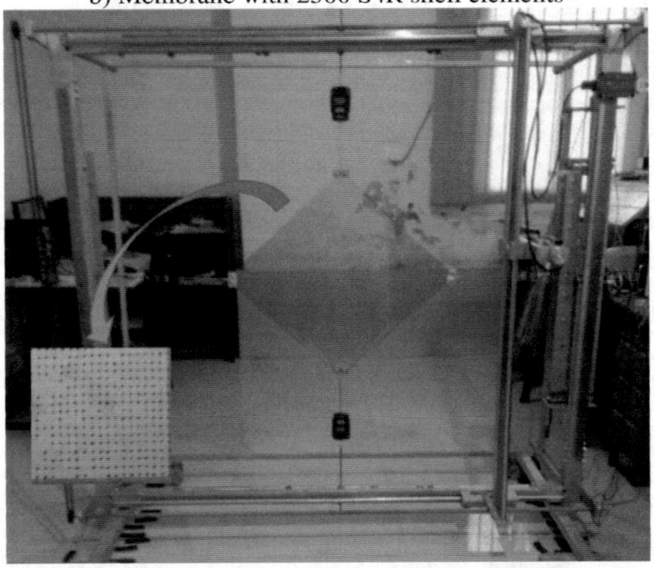

Figure 2. Experimental Analysis of Square Membrane with tensile loading

5. Stress Analysis

The large out-of-plane deformations formed as local–post buckling pattern considered as wrinkles in membrane. Negligible bending stiffness and non-resistance of compressive stress in the membrane structures are main cause of the appearance of wrinkles [13]. Membrane structures will deflect immediately after occurrence of local in-plane compression within the structures and hence retain the in-plane principal stresses non-negative [15]. So, the discussion of the membrane mechanical has been distinct in two stages: State of stress before and after wrinkling.

The State of stress before wrinkling, the membrane is in flat conditions, i.e. state of plan stress. Here, membrane is considered as two-dimensional, isotropic elastic materials. Assuming that deformation is very small and body forces are zero before occurrence of wrinkle in the membrane

Both membrane deformations and bending deformation can be taken in to account after the wrinkle formation in membrane. The FEM modelling of the wrinkling phenomenon has highly non-linear in nature(Lu & Tan, 2012; Nayyar, Ravi-Chandar, & Huang, 2014;). The prediction of wrinkling formation in membrane will play a great role in numerical simulation and design procedure. There is three criteria to predict wrinkle formation: principal stress, principal strain and principal stress-strain criterion (Wang, Tan, Du, & Wan, 2007). In all three criterion, principal stress criterion has been chosen for the analysis.

Let us consider σ_1 and σ_2 are the major principal stress and minor principal stress respectively. If both major and minor principal stress are greater than zero, then membrane is in taut state. If major principal stress is greater than zero and minor principal stresses is non-positive then membrane is in wrinkled state. If both major and minor principal stresses are non-positive then membrane is in slack state.

Fig. 3 shows the distribution of major plane principal stresses at four locations of in the membrane. The stress limits are set at 0 to 30 N/mm^2 in order to better visualize the stress variation. The general trend is that the stress decreases as one moves away from the corners of the membrane. For T1/T2 =1 the higher stresses are localized near the four corners (see Fig. 4 (a)), but for increasing T1/T2 the major principal stresses tend to spread along the main diagonal (see Fig. 4 (b)). Because wrinkling is associated with the existence of (small) compressive stresses, it is instructive to consider also the distribution of the minor principal stresses.

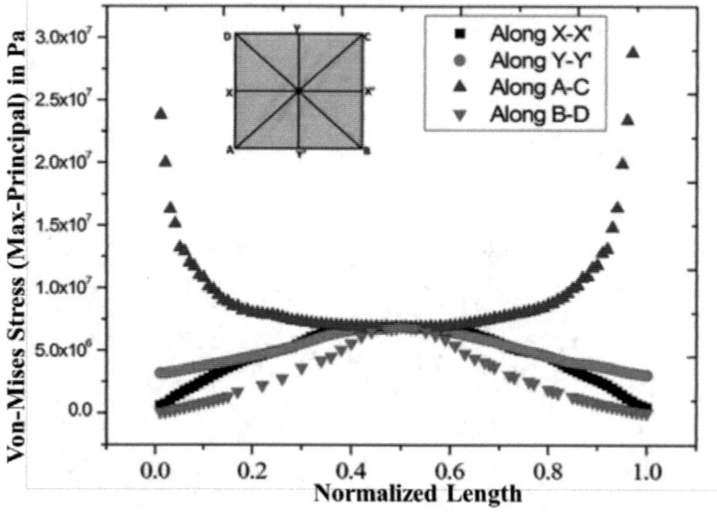

Figure 3. Von-Mises (minimal principal) Stress Value at different zone of the membrane structures

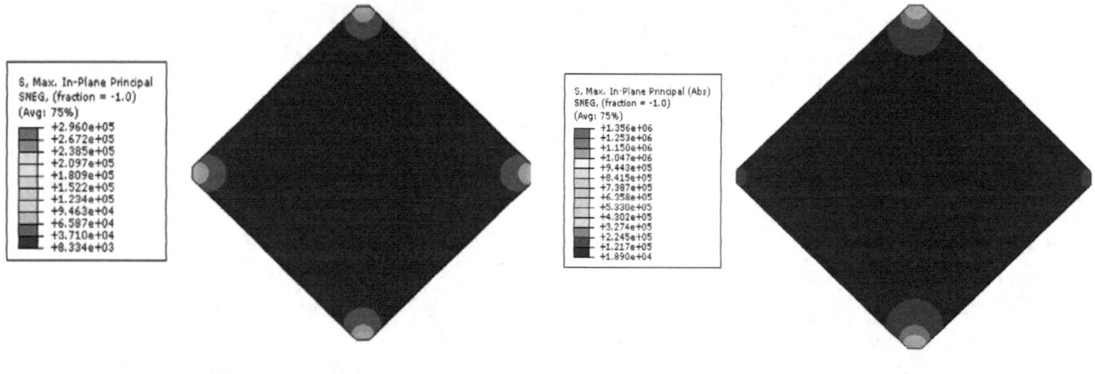

(a) $T_1/T_2=1$ (b) $T_1/T_2=5$

Figure 4. Stress distribution on membrane under different T_1/T_2 ration

6. Conclusion

In this paper, different structural and thermomechanical condition has been discussed to analyses wrinkling behavior of membrane structures. The following conclusions have been summarized.

The results confirmed that square membrane with corner tensile loading has four fans of wrinkles near the corners if $T_1/T_2<1/(\sqrt{2}-1)$. But they go through the center if $T_1/T_2 \geq 1/(\sqrt{2}-1)$ and less number of wrinkles will be formed along diagonal [24]. In this, it is also observed that surface error of the membrane structures increases with increases with T_1/T_2.

A comprehensive study of this issue needs to be continued. In the next article, the authors will consider a detailed analysis of the research.

7. Acknowledgement

The authors are truly thankful to Mr. Arup Kumar Hait, Group Head MAMG, Mr. Sammir S Sakhare, Hear AMDD, and Mr. Kripa Shankar Singh, Scientist/Engineer-SE, Space Applications Centre (ISRO) for their valuable and significant suggestions, which substantially improved the manuscript. This work supported by the Indian Space Research Organization, Government of India (Project Grant Number ISR-833-MID).

8. References

[1] C. H. M. Jenkins, "Recent Advances in Gossamer Spacecraft," *Progress in astronautics and aeronautics*, vol. 212. American Institute of Aeronautics and Astronautics, 2006.

[2] C. H. M. Jenkins, *Gossamer Spacecraft: Membrane and inflatable structures technology for space applications*, vol. 191. AIAA, 2001.

[3] E. J. Ruggiero and D. J. Inman, "Gossamer Spacecraft: Recent Trends in Design, Analysis, Experimentation, and Control," *J. Spacecr. Rockets*, vol. 43, no. 1, pp. 10–24, 2006.

[4] J. Santiago-Prowald and H. Baier, "Advances in deployable structures and surfaces for large apertures in space," *CEAS Sp. J.*, vol. 5, no. 3–4, pp. 89–115, 2013.

[5] L. Scialino, A. Ihle, M. Migliorelli, N. Gatti, L. Datashvili, K. van 't Klooster, and J. Santiago Prowald, "Large deployable reflectors for telecom and earth observation applications," *CEAS Sp. J.*, vol. 5, no. 3–4, pp. 125–146, 2013.

[6] R. E. Freeland and G. Bilyeu, "In-Step Inflatable Antenna Experiment," *Acta Astronaut.*, vol. 30, pp. 29–40, 1993.

[7] R. E. Freeland and G. Bilyeu, "In-Step Inflatable Antenna Experiment," *Int. Astronaut. Fed.*, 1992.

[8] O. Mori, H. Sawada, R. Funase, M. Morimoto, T. Endo, T. Yamamoto, Y. Tsuda, Y. Kawakatsu, J. Kawaguchi, and Y. Miyazaki, "IKAROS Demonstration Team, and Solar Sail Working Group,'First Solar Power Sail Demonstration by IKAROS.,'" *Trans. Japan Soc. Aeronaut. Sp. Sci. Aerosp. Technol. Japan*, vol. 8, 2009.

[9] A. Babuscia, T. Choi, J. Sauder, A. Chandra, and J. Thangavelautham, "Inflatable antenna for CubeSats: Development of the X-band prototype," in *IEEE Aerospace Conference Proceedings*, 2016, vol. 2016–June, pp. 1–11.

[10] N. Chahat, S. Member, R. E. Hodges, S. Member, J. Sauder, M. Thomson, E. Peral, and Y. Rahmat-samii, "CubeSat Deployable Ka-Band Mesh Reflector Antenna Development for Earth Science Missions," *IEEE Trans. Antennas Propag.*, vol. 64, no. 6, pp. 2083–2093, 2016.

[11] Y. Xu, F. Guan, H. Huang, and Q. Ye, "Thermal Distortion Analysis of Inflatable Antenna Structures Considering Inflation Gas," *J. Aerosp. Technol. Manag.*, vol. 8, no. 4, pp. 475–482, 2016.

[12] E. Cerda and L. Mahadevan, "Geometry and physics of wrinkling," *Phys. Rev. Lett.*, vol. 90, no. 7, p. 74302, 2003.

[13] E. Cerda, K. Ravi-Chandar, and L. Mahadevan, "Thin films: Wrinkling of an elastic sheet under tension," *Nature*, vol. 419, no. 6907, p. 579, 2002.

[14] M. Liu, J. Huang, and Y. Wang, "Analysis of wrinkled membrane structures based on a wrinkle-wave model," *AIP Adv.*, vol. 7, no. 1, 2017.

[15] K. Efimenko, M. Rackaitis, E. Manias, A. Vaziri, L. Mahadevan, and J. Genzer, "Nested self-similar wrinkling patterns in skins," *Nat. Mater.*, vol. 4, no. 4, p. 293, 2005.

[16] H. Soni, R. A. Pelcovits, and T. R. Powers, "Wrinkling of a thin film on a nematic liquid-crystal elastomer," *Phys. Rev. E*, vol. 94, no. 1, 2016.

[17] C. W. J. Oomens and J. D. Janssen, "The wrinkling of thin membranes: Part I—theory," *J. Appl. Mech.*, vol. 54, p. 885, 1987.

[18] D. G. Roddeman, J. Drukker, C. W. J. Oomens, and J. D. Janssen, "The wrinkling of thin

membranes: Part II—numerical analysis," *J. Appl. Mech.*, vol. 54, no. 4, pp. 888–892, 1987.

[19] W. Wong and S. Pellegrino, "Wrinkled Membranes III: Numerical Simulations," *J. Mech. Mater. Struct.*, vol. 1, no. 1, pp. 63–95, 2006.

[20] Q. Tao, C. Wang, Z. Xue, Z. Xie, and H. Tan, "Wrinkling and collapse of mesh reinforced membrane inflated beam under bending," *Acta Astronaut.*, vol. 128, pp. 551–559, 2016.

[21] R. Zhao and D. Wei, "Wrinkling Analysis of Stiffness Equivalence in Membrane Structures," in *Advances in Structures, Pts 1-5*, vol. 163–167, L. J. Li, Ed. 2011, pp. 1976–1979.

[22] C. G. Wang, H. F. Tan, X. W. Du, and Z. M. Wan, "Wrinkling prediction of rectangular shell-membrane under transverse in-plane displacement," *Int. J. Solids Struct.*, vol. 44, no. 20, pp. 6507–6516, 2007.

[23] Y. W. Wong and S. Pellegrino, "WRINKLED MEMBRANES PART III : NUMERICAL SIMULATIONS," *Mech. Mater. Struct.*, vol. 1, no. 1, 2006.

[24] Y. W. Wong and S. Pellegrino, "WRINKLED MEMBRANES PART II : ANALYTICAL MODELS," *Mech. Mater. Struct.*, vol. 1, no. 1, 2006.

[25] Y. W. Wong and S. Pellegrino, "WRINKLED MEMBRANES PART I : EXPERIMENTS," *J. Mech. Mater. Struct.*, vol. 1, no. January, pp. 61–93, 2006.

[26] R. K. Miller and J. M. Hedgepeth, "An algorithm for finite element analysis of partly wrinkled membranes," *AIAA J.*, vol. 20, no. 12, pp. 1761–1763, 1982.

[27] M. Stein and J. M. Hedgepeth, "Analysis of partly wrinkled membranes," National Aeronautics and Space Administration Washington, 1961.

[28] H. Furuya, Y. Miyazaki, and Y. Akutsu, "Experiments on Static Shape Control of One-Dimensional Creased Membrane By Piezoelectric Films," in *43rd AIAA/ASME/ASCE/AHS/ASC Structures, Structural Dynamics, and Materials Conference*, 2002, no. April, pp. 1–4.

[29] N. Damil, M. Potier-Ferry, and H. Hu, "Membrane wrinkling revisited from a multi-scale point of view," *Adv. Model. Simul. Eng. Sci.*, vol. 1, no. 1, p. 6, 2014.

[30] J. R. Blandino, J. D. Johnston, and U. K. Dharamsi, "Corner wrinkling of a square membrane due to symmetric mechanical loads," *J. Spacecr. Rockets*, vol. 39, no. 5, pp. 717–724, 2002.

[31] K. Attipou, H. Hu, F. Mohri, M. Potier-Ferry, and S. Belouettar, "Thermal wrinkling of thin membranes using a Fourier-related double scale approach," *Thin-Walled Struct.*, vol. 94, pp. 532–544, 2015.

[32] Q. Huang, H. Hu, K. Yu, M. Potier-Ferry, S. Belouettar, and N. Damil, "Macroscopic simulation of membrane wrinkling for various loading cases," *Int. J. Solids Struct.*, vol. 64, pp. 246–258, 2015.

[33] E. Lejeune, A. Javili, and C. Linder, "An algorithmic approach to multi-layer wrinkling," *Extrem. Mech. Lett.*, vol. 7, pp. 10–17, 2016.

[34] T. C. Jones, H. Bart-Smith, M. M. Mikulas, and J. J. Watson, "Finite Element Modeling and Analysis of Large Pretensioned Space Structures," *J. Spacecr. Rockets*, vol. 44, no. 1, pp. 183–193, 2007.

Determination of pervious concrete mix by strength

Bo-Syun Huang[1], Yan-Fu Li[1], Chi-Hsing Wu[2] and Chi Tang[3*]

1 Graduate Student, Department of Soil and Water Conservation, National Pingtung University of Science and Technology, Pingtung, Taiwan
2 Associate Professor, Department of Civil Engineering, National Pingtung University of Science and Technology, Pingtung, Taiwan
3 Corresponding author: Associate Professor, Department of Soil and Water Conservation, National Pingtung University of Science and Technology, Pingtung, Taiwan

*E-mail: tangchi@mail.npust.edu.tw

Abstract. The paper first used a cement amount *C*=300 kg and a range of water-cement ratio *W/C*=0.36~0.44 to make pervious concrete specimens of different *W/C* to perform the compressive strength and the flexural strength tests, as well as the permeability coefficient and the porosity tests. These tests were to find the *W/C* of the maximum strength under meeting the requirements of permeability and water retention. The concrete mix corresponding to the *W/C* determined this way can be treated as the optimal mix since it gives the maximum strength for a cement amount *C*=300 kg. Then, similar tests were performed for *C*= 350 kg to determine its optimal mix. From which, the relationship of the strength of the optimal mix vs *C* is established.

1. Introduction

Traditional road pavements are paved by impervious material pavement, which will increase impervious area, add surface runoff and lead to flood problem. The water absorption and retention capability of soil is reduced due to lacking of seeping path of rainwater, and result in *heat island effect* in urban area. Additionally, the puddles on the road usually cause vehicles skid when raining. It also causes water mist and splashing when vehicles running, affects the safety of pedestrians and driving. If the traditional road pavements are replaced as pervious pavement, it can not only improve the urban flooding and *heat island effect* problems, but also enhance the safety of pedestrians and driving in raining days. However, roadbed of pervious pavement will become soft due to rain infiltration. If the bearing capacity of base and bottom courses is not enough, it often causes the road collapse to fail in high vehicle load. For this reason, US California Department of Transportation (Caltrans) in its pervious pavement design guidance [1] does not propose the pervious pavement used for road condition of Category D & E, which are categories of moderate are high heavy vehicular loads with high speed respectively.

Keda road pervious pavement in Neipu Township, Pingtung county, Taiwan was completed on July 29, 2018 (2018/7/29). It is four-lane road with total length 3,275 m. The lane width of north or south bounds is 8.8 m. The road was designed for high heavy and high speed vehicular load, and has been used for more than 14 months so far. The pavement offers very good traffic service and no pothole or collapse problems. The key to the success of Keda road pervious pavement is the pervious concrete in the base course is strong enough to resist vehicle load even the roadbed become soft by infiltrating rain. No wonder, pervious concrete has been recognized by US Environmental Protection Agency (EPA) as a Best Management Practices (BMP) for storm water management. Pervious concrete pavement may be an unique and effective means to address important environmental issues and support green, sustainable growth. Its technology creates more efficient land use by eliminating the need for retention ponds, swales, and other storm water devices. Thus, pervious concrete has the ability to lower overall project on a first-cost basis.

Pervious concrete is a mixture of cement, water, and coarse aggregate, and little to no sand. It creates a very porous medium that allows water to drain to the underlying soils. Commonly, it may allow as much as 5000 mm/hr to pass through the body of the concrete. However, the pervious concrete usually gives a low strength due to inappropriate mix and improper construction.

The paper establishes the curves of strength vs water-cement ratio *W/C* to determine the concrete mix. The mix corresponding to the maximum strength for a given cement amount *C* can be treated as the optimal mix since it gives the possible largest strength effect for a cement amount without adding any chemical admixture. It is the most economical mix due to that the cement is the most expansive material in pervious concrete compositions. The strength tests include the compressive strength test and the flexural strength test. Normally, the former is used to determine the optimal mix. The permeability coefficient test and the porosity test were performed as well. The former is to investigate the permeability of pervious concrete, in which the permeability coefficient *k* is required larger than 10^{-3} cm/sec according to the Outline Specification for Public Construction [2] by Public Construction Commission, or more strictly required larger than 10^{-2} cm/sec according to the Evaluation Manual for Green Building Material [3] by Taiwan Architecture & Building Center. The latter is to investigate the water retention ability of pervious concrete, in which the porosity *n* is required larger than 15% according to the Evaluation Manual for Green Building Material [3].

The paper used cement amounts *C*=300 and 350 kg to perform mix design and tests respectively. The optimal mixes are then determined by the compressive strength test results for *C*=300 and 350 kg respectively. From which, the relationship of the strength of the optimal mix vs *C* is established.

2. Test plan

The tests are divided into three parts: coarse aggregate material property tests, fresh concrete tests, and pervious concrete specimen tests. They are described as follows.

2.1. Coarse aggregate material property tests

Coarse aggregate material property tests have the sieve test and the specific gravity and absorption test. They are described as follows.

(1) Sieve analysis test: According to CNS 486 "Method of test for sieve analysis of fine and coarse aggregates" [4], the aggregate sieve analysis test is performed to plot grain size distribution curve, and to gain the fine modulus and the grain size distribution.

(2) Specific gravity and absorption test: According to CNS 488 "Method of test for density, relative density (specific gravity), and absorption of coarse aggregate" [5], the specific gravity, absorption and water content of aggregate are determined. The oven-dry specific gravity will be used in the calculation of pervious concrete mix design, and the absorption and water content of aggregate will be used for adjusting the add water amount.

2.2. Fresh concrete tests

Fresh concrete tests include the drooping flow test and the unit weight and air content test. They are described as follows.

(1) Drooping flow test: Using the net cage method proposed by Hu [6] and Pan [7], the drooping flow amount of fresh concrete on is weighted to plot the droop flow curve of drooping flow amount vs *W/C* to estimate the possible *W/C* range of maximum strength, which will be used for selecting the *W/C* of specimens made for the pervious concrete test to save time and cost of the tests.

(2) Unit weight and air content test: According to CNS 11151 "Method of test for unit weight, yield, and air content (gravimetric) of concrete" [8], the fresh concrete is tested to find its unit weight and air content.

2.3. Pervious concrete specimen tests

Pervious concrete specimen tests include the compressive strength and the flexural strength tests, the permeability coefficient and the porosity tests. The former two tests are performed to evaluate the safety of pervious concrete. The latter two tests are performed to evaluate the permeability and water retentivity of pervious concrete respectively. They are described as follows.

(1) Compressive strength test: According to CNS 1232 "Method of test for compressive strength of cylindrical concrete specimen: [9], the compressive strength of concrete specimen is measured.

(2) Flexural strength test: According to CNS 1233 "Method of test for flexural strength of concrete (using simple beam with third-point loading)" [10], the modulus of rupture of concrete specimen is measured. The modulus of rupture is a measurement for the concrete tensile strength.

(3) Permeability coefficient test: According to CNS 14995 "Permeable concrete paving blocks" [11], the permeability coefficient of concrete specimen is measured.

(4) Porosity test: Referring to the test steps of CNS 488 [5], the porosity of concrete specimen can be evaluated by a proposed formula.

3. Methods of making specimens and samples

The size of specimen or sample for each test is listed in Table 1. The tool to impact the specimen or sample is steel impactor. The impactor has large impactor and small impactor. The detailed sizes of impactors are shown in Figure 1. The large impactor is designed for the ψ15×30 cm cylindrical specimen of the compressive strength test, bust also can be used for making the specimen of the flexural strength test, for the samples of the drooping flow test and the unit weight and air content test of fresh concrete. The small impactor is designed for the ψ10×20 cm cylindrical specimen of the permeability coefficient test. The porosity test can use the same specimen as that of the permeability coefficient test. The design parameters of the impactors are listed in Table 2. The impactor has been obtained No. I628335 invention patent by the Intellectual Property Office, Ministry of Economic Affair, ROC. Figure 2 shows the impacting operation using an impactor.

The fresh concrete is filled into the mold or container by several layers when making specimen or

Table 1. Size, number of layers and impact times of each specimen and sample

Concrete Phase type	Test name	Size of specimen or sample (cm)	Impact tool	No of layer	Layer height (cm)	Impact times each layer
Fresh Concrete	Drooping flow test	30×16×12	Large impactor	2	6	34
	Unit weight and air content test	ψ20.3×21.7	Large impactor	4	5.4	24
Harden Concrete	Compressive strength test	ψ15×30	Large impactor	6	5	12
	Flexural strength test	53×15×15	Large impactor	3	5	54
	Permeability coefficient test Porosity test	ψ10×20	Small impactor	2	5	12

Table 2. Parameters of impactors

Parameters	Large impactor	Small impactor
Specimen size (cm)	ψ15×30	ψ10×20
Hammer size (cm)	ψ8×13.0	ψ8×6.0
Hammer mass (kg)	4.48	2.02
Falling Distance (cm)	54	54
Unit momentum per impact (N · sec/cm^2)	0.0821	0.0825

impacted several time. The point of numbers of layers and impact times is to produce equivalent compactness for all the specimens and samples, and these specimens and samples have the similar

IOP Conf. Series: Materials Science and Engineering **812** (2020) 012015 doi:10.1088/1757-899X/812/1/012015

compactness of the pervious concrete pavement in-situ. Thus test results using these specimens or samples can actually reflect and simulate the pervious concrete pavement in-situ. The number of layers and impact times for each specimen and sample is also listed in Table 1.

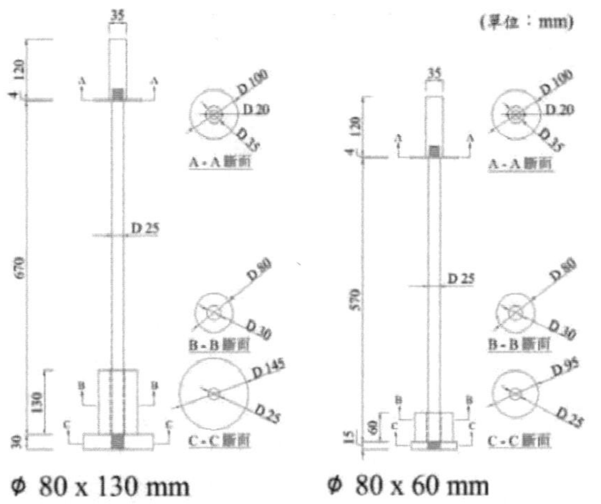

Figure 1. Sizes of impactors

Figure 2. Impacting operation

Figure 3. Coarse aggregate screen

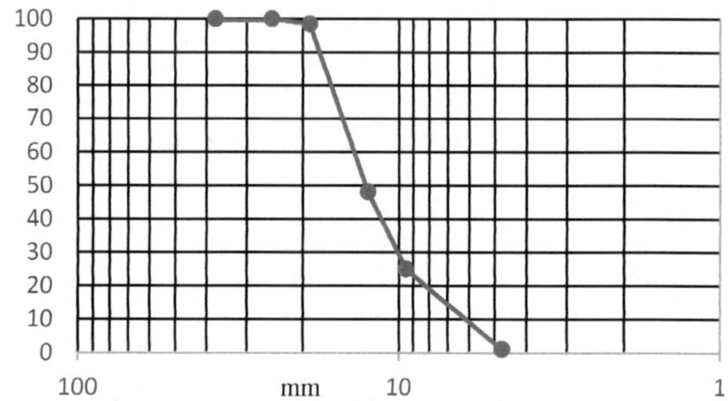

Figure 4. Grain size distribution curve

4. Results and discussions

The test results are divided into three parts: coarse aggregate material property tests, fresh concrete tests, and pervious concrete specimen tests. They are described as follow

4.1. Coarse aggregate material property test results and discussions

Coarse aggregate material property tests include the sieve test and the specific gravity and absorption test. The test results are discussed as follows.

(1) Sieve analysis test: According to CNS 486, 2000 g coarse aggregate was taken to perform the aggregate sieve analysis test. Figure 3 is the coarse aggregate screen for the sieve analysis test. The test results are listed in Table 3. The table shows that, the maximum size of the aggregate is 25.0 mm. The fineness modulus *FM* is 6.76. Average grain size is 12.5 mm. Using the passing percentage in Table 3 to plot curve vs the aggregate size, a grain size distribution curve shown as Figure 4 is

obtained. The curve shows that, the aggregate for the test is basically uniform graded. The main grain size range is in #4~3/4″ (4.75~19.0 mm).

Table 3. Sieve analysis test results of aggregate

Sieve no (mm)	Sieve weight (g)	(Sieve+ Retained) weight (g)	Retained weight (g)	Retained percentage (%)	Cumulative retained percentage (%)	Passing percentage (%)
$1\frac{1}{2}$″ (37.5)	-	-	-	0.0	0.0	100.0
1″ (25.0)*	4745	4745	0	0.0	0.0*	100.0
$\frac{3}{4}$″ (19.0)	5375	5407	32	1.6	1.6	98.4
$\frac{1}{2}$″ (12.5)*	4802	5796	994	50.2	51.8*	48.1
$\frac{3}{8}$″ (9.5)	4993	5451	458	23.2	75.0	25.0
#4 (4.75)	5227	5701	474	24.0	99.0	1.0
Chassis	4577	4597	20	1.0	100.0*	0.0
Total			1978	100.0	175.6[a]	
FM [a]	6.76				Average grain size	12.5[b]

[a] *FM*=(cumulative retained percentage+500)/100. Half-sieves and chassis are not counted in cumulative percentage. Sieve no 1″ and $\frac{1}{2}$″ are half-sieves. Since #8~#100 have 5 sieves, *FM* must add 500.

[b] Average grain size= $0.016(\frac{25.0+19.5}{2})+0.502(\frac{19.0+12.5}{2})+0.232(\frac{12.5+9.5}{2})+0.240(\frac{9.5+4.75}{2})$
=0.36+7.91+2.55+1.71=12.5 mm

(2) Specific gravity and absorption test: According to CNS 488, 3000 g coarse aggregate was taken to perform the specific gravity and absorption test. Measure 3 to average. Figure 5 shows the operation of aggregate weighting in water in the test. Table 4 is the test results. The table shows that, the oven-dry (OD), the saturated surface-dry (SSD) and the apparent specific gravities are 2.54, 2.58 and 2.65 respectively, the OD specific gravity will be used in the calculation of pervious concrete mix design. The absorption and water content are 0.84% and 1.56% respectively, which will be used for adjusting the add water amount.

The mix design procedure for pervious concrete follows those of Hu [6] and S-J Pan [7]. Using the OD specific gravity, absorption and water content of coarse aggregate obtained above, the concrete mix proportion can be calculated as in Table 5. The reason why the range of *W/C* in the table is taken to be 0.36~0.44 for both *C*=300 kg and 350 kg will be explained in the drooping flow test in the next subsection.

4.2. *Fresh concrete test results and discussions*

Fresh concrete tests include the drooping flow test and the unit weight and air content test. The test results are discussed as follows.

(1) Drooping flow test: The drooping flow test results are listed in Table 6. Using the drooping flow amount vs *W/C* in the table to plot curves, the droop flow curves shown as Figure 6 are obtained. The intersections of the horizontal and rising tangent lines are read at *W/C*=0.38 and 0.40 for *C*=300 kg

and 350 kg respectively. The maximum strength would be near these *W/C* values. Therefore, the test range could be chosen in *W/C*=0.36~0.44 for both *C*=300 and 350 kg.

Table 4. Specific gravity and absorption test results of aggregate

Items	Test values			
	1	2	3	Average
(1)Sample weight W_1 (g)	3000	3000	3000	3000
(2)Plate weight (g)	88	87	86	87
(3)Oven-dry (OD) (sample+plate) weight (g)	3066	3065	3055	3062
(4)OD sample weight W_2 (g)	2978	2978	2969	2975
(5)Water content $W_w = W_1 - W_2$ (g)	22	22	31	25
(6)Water content ratio $M = \frac{W_w}{W_2} \times 100\%$	0.74	0.74	1.04	0.84
(7)Saturate surface-dry(SSD) weight W_3 (g)	3003	2999	3020	3007
(8)Cage weight in water W_4 (g)	493	493	493	493
(9)(Cage+sample) weight in water W_5 (g)	2341	2324	2341	2335
(10)Apparent mass of sample in water immersion $E = W_5 - W_4$ (g)	1848	1831	1848	1842
(11)Water weight of the same volume as SSD sample= $W_3 - E$ (g)	1155	1168	1172	1165
(12)OD weight in water immersion W_6 (g)	2960	2958	2965	2961
(13)Water absorption ratio $M_S = \frac{W_3 - W_6}{W_6} \times 100\%$	1.45	1.39	1.85	1.56
(14)OD specific $S_d = \frac{W_6}{v_g}$	2.56	2.53	2.53	2.54
(15)SSD specific $S_s = \frac{W_3}{v_g}$	2.60	2.57	2.58	2.58
(16)Apparent specific $S_a = \frac{W_6}{W_6 - E}$	2.66	2.62	2.65	2.65

(2) Unit weight and air content test: The unit weight and air content test results are listed in Table 7. Using the unit weight *U* and the air content *Air* vs *W/C* in the table to plot line graphs, the results are shown as Figure 7 and 8 respectively. The figures show that, the *U* of *C*=350 kg is larger than that of *C*=300 kg, while the *Air* is smaller at the same *W/C*. That means that, more cement amount *C* will produce more slurry in fresh concrete such that its *U* is larger and *Air* is smaller for a same *W/C*. Larger *U* of larger *C* implies that, it will give a more compact density of pervious concrete after concrete become harden. Thus its strength will be higher and permeability and porosity be lower. Other thing to be mention is that, both air content curves in Figure 8 show an obvious drop at *W/C*=0.40. It may be that, the slurry become obvious increase after *W/C*=0.40 such that the *Air* begin to obvious drop down. The results are very coincident with those of the drooping flow test, where the slurry begin obviously getting more at *W/C*=0.38 for *C*=300 kg and at *W/C*=0.40 for *C*=350 kg respectively.

Figure 5. Specific gravity test of coarse aggregate

Table 5. Pervious concrete mix

Cement amount *C* (kg)	Water-cement ratio *W/C*	Assumed air content *Air* (%)	Water Amount (kg) Trial mix	Water Amount (kg) Field mix	Aggregate [a] amount (kg) Trial mix	Aggregate [a] amount (kg) Field mix	Total material mass (kg)
300	0.36	21.0	108	119	1490	1503	1922
	0.38	20.5	114	125	1488	1500	1925
	0.40	20.0	120	131	1485	1498	1928
	0.42	19.5	126	137	1483	1495	1932
	0.44	19.0	132	143	1480	1493	1935
350	0.36	21.0	126	136	1404	1416	1902
	0.38	20.5	133	143	1399	1411	1904
	0.40	20.0	140	150	1394	1406	1906
	0.42	19.5	147	157	1389	1401	1908
	0.44	19.0	154	164	1384	1396	1910

[a] Aggregate: OD specific gravity = 2.54, SSD specific gravity= 2.58, water content ratio = 0.84%, absorption = 1.56%.

Table 6. Drooping flow test results

Cement amount *C* (kg)	Water-cement ratio *W/C*	(Plate[a]+ slurry) weight (g)	Drooping flow amount (g)
300	0.36	395	14
	0.38	404	23
	0.40	585	204
	0.42	599	218
	0.44	602	221
350	0.36	408	27
	0.38	523	142
	0.40	528	147
	0.42	731	350
	0.44	957	576

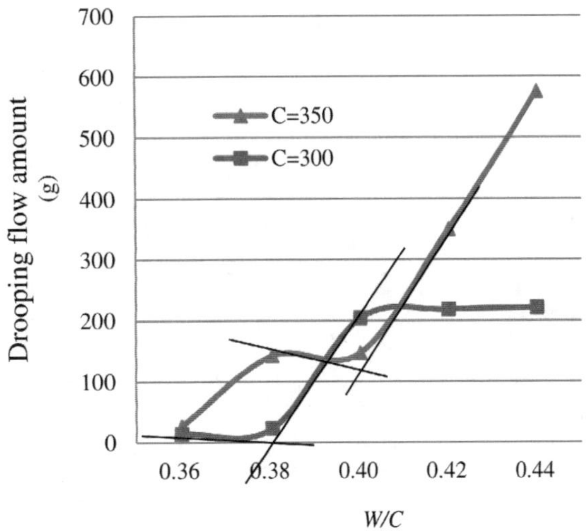

Figure 6. Drooping flow curves

Table 7. Unit weight and air content test results

Cement amount C (kg)	Water-cement ratio W/C	Assumed air content Air (%)	Total material mass W_1 (kg)	(Barrel [a] +mixed concrete) weight (kg)	Mixed concrete weight (kg)	Measured unit weight U [b] (kg/m³)	Measured concrete volume V [c] (m³)	Total material volume V_1 [d] (m³)	Measured air content Air [e] (%)	△Air (%)
300	0.36	21.0	1922	16.417	13.825	1982	0.969	0.790	18.5	-2.5
	0.38	20.5	1925	16.373	13.781	1976	0.974	0.795	18.4	-2.1
	0.40	20.0	1928	16.351	13.759	1973	0.977	0.800	18.2	-1.8
	0.42	19.5	1932	16.419	13.827	1983	0.974	0.805	17.4	-2.1
	0.44	19.0	1935	16.571	13.979	2004	0.966	0.810	16.1	-2.9
350	0.36	21.0	1902	16.597	14.005	2008	0.947	0.790	16.6	-4.4
	0.38	20.5	1904	16.421	13.829	1983	0.960	0.975	17.2	-3.3
	0.40	20.0	1906	16.372	13.780	1976	0.965	0.800	17.1	-2.9
	0.42	19.5	1908	16.525	13.933	1998	0.955	0.805	15.7	-3.8
	0.44	19.0	1910	16.566	13.974	2004	0.953	0.810	15.0	-4.0

[a] barrel inner size:ψ20.3×21.7; barrel weight=2.592 kg; content volume of barrel=6.974 L= 0.06974 m³
[b] measured U=mixed concrete weight (kg)/ 0.06974 m³
[c] measured concrete $V=W_1/U$
[d] total material volume V_1=(1-assumed Air)(1m³)
[e] measured $Air(\%)=(V-V_1)/V\times100\%$

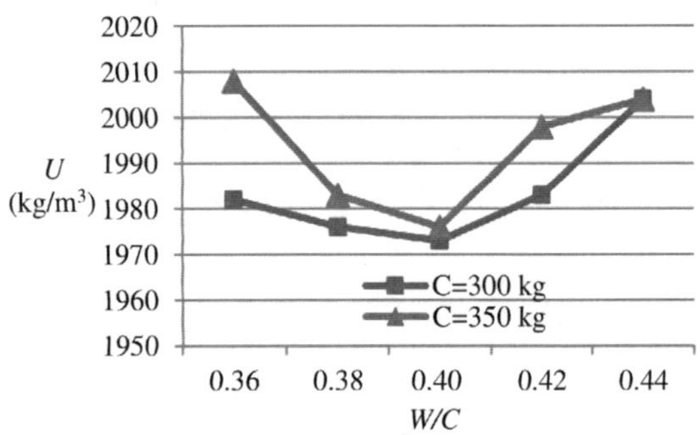

Figure 7. Line graphs of measured U vs water-cement ratio W/C

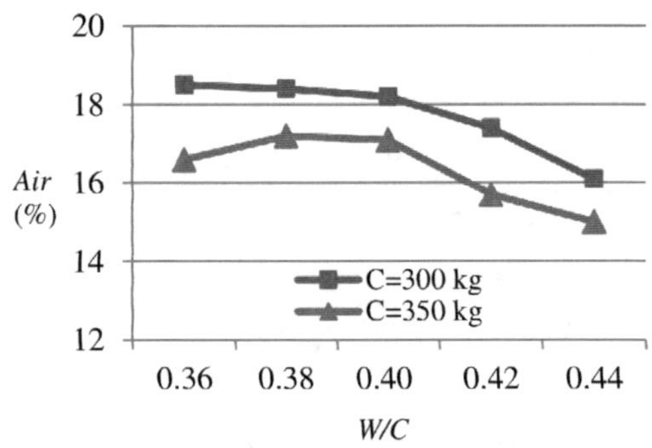

Figure 8. Line graphs of measured Air vs water-cement ratio W/C

4.3. Pervious concrete specimen test results and discussions

Pervious concrete specimen tests include the compressive strength and the flexural strength tests, the permeability coefficient and the porosity tests. The test results are discussed as follows.

(1) Compressive strength test: The compressive strength test results are listed in Table 8. The table shows that, the ratio of the 7th day to 28th day compressive strength f_{c7}/f_{c28} is in the range of 0.70~0.87, average 0.79 for C=300 kg. The range of f_{c7}/f_{c28} is 0.80~0.87, average 0.84 for C=350 kg. Using the compressive strength f_c in the table vs W/C to plot line graphs, the results are shown as Figure 9. The figure shows that, both the maximum f_c of the 7th and 28th days are at W/C=0.38 for C=300 kg. The values are 159 and 194 kgf/cm^2 (15.6 and 19.0 MPa) respectively. Thus, the ratio f_{c7}/f_{c28} =0.82 at W/C=0.38. Both the maximum f_c of the 7th and 28th days are at W/C=0.40 for C=350 kg. The values are 212 and 251 kgf/cm^2 (20.8 and 24.6 MPa) respectively. Thus, the ratio f_{c7}/f_{c28} =0.84 at W/C=0.40.

(2) Flexural strength test: The flexural strength test results are also listed in Table 8. The table shows that, the ratio of the 7th day to 28th day modulus of rupture R_7/R_{28} is in the range of 0.67~0.79, average 0.74 for C=300 kg. The ratio of the 28th day modulus of rupture to 28th day compressive strength R_{28}/f_{c28} is in the range of 0.16~0.22, average 0.19 for C=300 kg. The ranges of R_7/R_{28} and R_{28}/f_{c28} are 0.74~0.89 and 0.13~0.20, average 0.83 and 0.16 respectively for C=350 kg. Using the modulus of rupture R in the table vs W/C to plot line graphs, the results are shown as Figure 10. The figure shows that, both the maximum R of the 7th day and 28th day are at W/C=0.38 for C=300 kg. The values are 22.6 and 31.4 kgf/cm^2 (2.2 and 3.1 MPa) respectively. Thus, the ratio R_7/R_{28} =0.67 and R_{28}/f_{c28}=0.16. Both the maximum R of 7th day and 28th day are at W/C=0.40 for C=350 kg. The values are 27.5 and 32.7 kgf/cm^2 (2.7 and 3.2 MPa) respectively. Thus, the ratio R_7/R_{28} =0.84 and R_{28}/f_{c28}=0.13. The W/C values of the maximum strength in the flexural strength test is exactly the same as those of the compressive strength test for C=300 and 350 kg.

Table 8. Compressive strength test and flexural strength test results

Cement amount C (kg)	Water-cement ratio W/C	Compressive strength f_c (kgf/cm^2) 7th day f_{c7} 1	2	3	Avg	28th day f_{c28} 1	2	3	Avg	f_{c7}/f_{c28}	Modulus of rupture R (kgf/cm^2) 7th day R_7	28th day R_{28}	R_7/R_{28}	R_{28}/f_{c28}
300	0.36	128	133	133	132	170	169	154	165	0.80	21.0	30.8	0.68	0.16
	0.38	157	158	161	159	221	176	185	194	0.82	22.6	31.4	0.67	0.16
	0.40	124	100	103	109	151	168	150	156	0.70	22.3	28.1	0.79	0.18
	0.42	118	108	119	115	136	125	137	132	0.87	22.0	28.5	0.77	0.22
	0.44	102	107	90	99	123	132	134	129	0.77	22.6	28.6	0.79	0.22
									Avg	0.79			0.74	0.19
350	0.36	122	137	130	130	119	155	171	148	0.87	24.5	29.1	0.84	0.20
	0.38	135	131	156	141	167	191	150	169	0.83	24.9	29.0	0.86	0.17
	0.40	224	197	215	212	247	260	247	251	0.84	27.5	32.7	0.84	0.13
	0.42	177	156	187	173	207	201	211	206	0.84	23.5	31.9	0.74	0.15
	0.44	136	171	127	144	178	180	185	181	0.80	24.5	27.5	0.89	0.15
									Avg	0.84			0.83	0.16

(3) Permeability coefficient test: The permeability coefficient test results are listed in Table 9. The table shows that, the permeability coefficient k is in the range of 0.070~0.185 cm/sec. All the k values are much larger than the requirement 10^{-2} cm/sec. Using the k in the table vs W/C to plot line graphs, the results are shown as Figure 11. The figure shows that, the minimum k values are at W/C=0.38 and 0.42 for C=300 and 350 kg respectively. The values are 0.083 and 0.070 cm/sec respectively. The

result is exactly the same as the *W/C*=0.38 of the maximum compressive strength for *C*=300 kg, and very close to the *W/C*=0.40 of the maximum compressive strength for *C*=350 kg. The high compactness of pervious concrete would give high strength, but low permeability. Therefore, the results of the permeability coefficient test are quite coincident with those of the compressive strength test.

(4) Porosity test: The porosity test results are also listed in Table 9. The table shows that, the porosity *n* is in the range of 36.3%~40.8%. All the *n* values are much larger than the requirement 15%. Using the *n* in the table vs *W/C* to plot line graphs, the results are shown as Figure 12. The figure shows that, the minimum *n* values are at *W/C*=0.38 and 0.42 for *C*=300 and 350 kg respectively. The values are 36.7% and 36.3% respectively. The results are exactly the same as the *W/C*=0.38 of the maximum compressive strength for *C*=300 kg, and very close to the *W/C*=0.40 of the maximum compressive strength for *C*=350 kg. The high compactness of pervious concrete would give high strength, but low porosity. Therefore, the results of the porosity test are quite coincident with that of the compressive strength test.

5. Further discussions on the test results

Comparison among the *W/C* values of the intersections of tangent lines on drooping flow curves, the obvious dropping points of air content line graphs, the maximum compressive strength, the maximum modulus of rupture, the minimum permeability coefficient, and the minimum porosity can give a table as Table 10. The table shows that, the *W/C* values of the maximum strength are totally coincident either for *C*=300 kg or 350 kg, for the compressive strength or the modulus of rupture, for the 7th or 28th day strengths. All maximum strengths are at *W/C*=0.38 for *C*=300 kg, at *W/C*=0.40 for *C*=350 kg. It can be sure of that, the optimal mix is at *W/C*=0.38 for *C*=300 kg, and at *W/C*=0.40 for *C*=350 kg from these results. Table 10 also shows that, the *W/C* values of drooping flow amount are totally coincident with those of the strength for C=300 kg and 350 kg. This result confirms that, the drooping flow test is able to estimate the possible W/C range of the maximum strength of pervious concrete.

Table 9. Permeability coefficient test and porosity test results

Cement amount *C* (kg)	Water-cement ratio *W/C*	Permeability coefficient *k* (cm/sec)	Porosity *n* (%)
300	0.36	0.106	39.8
	0.38	0.083	36.7
	0.40	0.179	37.1
	0.42	0.118	39.4
	0.44	0.176	39.9
350	0.36	0.185	40.8
	0.38	0.164	38.5
	0.40	0.110	36.5
	0.42	0.070	36.3
	0.44	0.108	36.5

Table 10. Comparison of water-cement ratio *W/C* of test results

Test name		Cement amount *C* (kg)	
		300	350
Drooping flow test		0.38	0.40
Unit weight and air content test		0.40	0.40
Compressive strength test	7th day	0.38	0.40
	28th day	0.38	0.40
Flexural strength test	7th day	0.38	0.40
	28th day	0.38	0.40
Permeability coefficient test		0.38	0.42
Porosity test		0.38	0.42

IOP Conf. Series: Materials Science and Engineering **812** (2020) 012015 doi:10.1088/1757-899X/812/1/012015

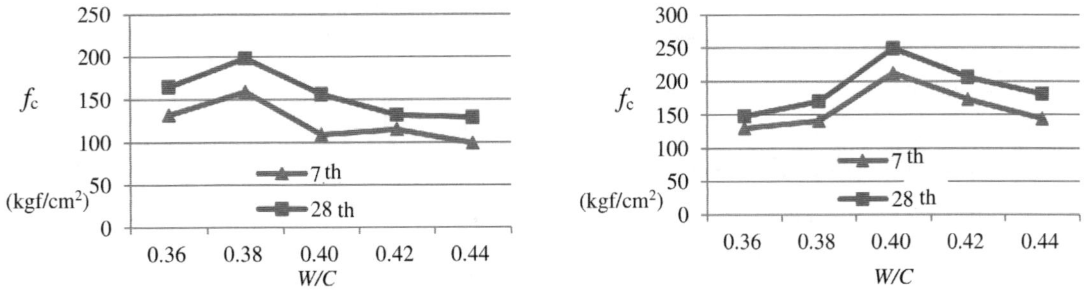

(A) C=300 kg (B) C=350 kg

Figure 9. Line graphs of compressive strength f_c vs water-cement ratio *W/C*

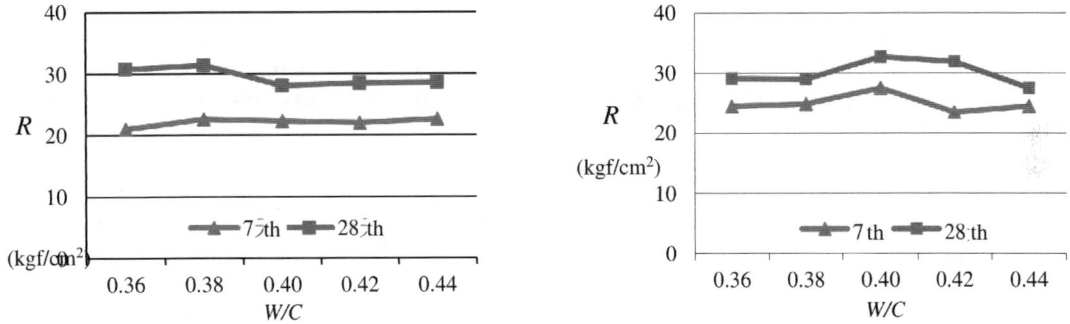

(A) C=300 kg (B) C=350 kg

Figure 10. Line graphs of modulus of rupture R vs water-cement ratio *W/C*

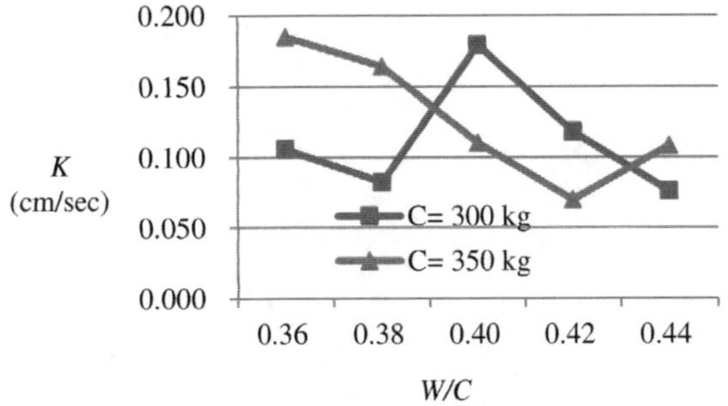

Figure 11. Line graphs of permeability coefficient k vs water-cement ratio *W/C*

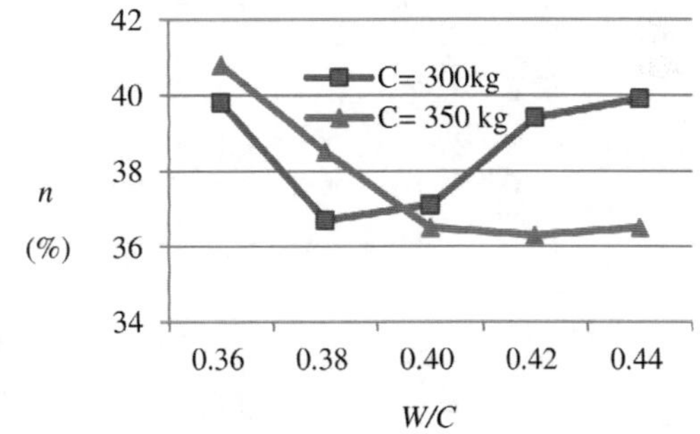

Figure 12. Line graphs of porosity n vs water-cement ratio *W/C*

For C=300 kg, only the W/C=0.40 of the air content is not coincident with W/C=0.38, but still very close to it. For C=350 kg, only the W/C=0.42 of the permeability coefficient and the porosity is not coincident with W/C=0.40, but also very close to it. Therefore, these test results provide evidences for the optimal mix determined above.

6. Revising the optimal mixes

A concrete mix proportion is defined as that the material amount of it will produce 1 m^3 volume of concrete after mixed. It is learned that from Table 7, for C=300 kg, the measured air content of

W/C=0.38 is 18.4%, smaller than the assumed air content 20.5%. Thus, the optimal mix in Table 5 will produce a volume of concrete smaller than 1 m^3. Therefore, the optimal mixes in Table 5 should be properly revised.

Assuming that the actual volume of the mix in Table 5 is *V* after mixed, *V* can be obtained by the equation

$$(1\text{-measured } Air)V = 1\text{-assumed } Air \qquad (1)$$

From Table 7, it is read that measured *Air*=18.4% and assumed *Air*=20.5% at *W/C*=0.38 for *C*=300 kg. Substituting them into Equation (1), then

$$(1-0.184)V = 1-0.205$$

gives *V*=0.974 m^3. From Table 7, it is read that measured *Air*=17.1% and assumed *Air*=20.0% at *W/C*=0.40 for *C*=350 kg. Substituting them into Equation (1), then

$$(1-0.171)V = 1-0.200$$

gives *V*=0.965 m^3. Dividing the mixes of *W/C*=0.38 for *C*=300 kg and *W/C*=0.40 for *C*=350 kg by 0.974 and 0.965 respectively, the revised mixes are obtained. The results are listed in Table 11, which are the optimal mixes finally determined for pervious concrete of *C*=300 and 350 kg.

From Table 11, the relationship between cement amount *C* and the 28th day compressive strength may be established by interpolating as

$$f_c = 1.04C - 125 \text{ kgf/cm}^2 \text{ for } C = 300\text{~}370 \text{ kg} \qquad (2)$$

This formula can be used to estimate the optimal compressive strength for a pervious concrete mix of cement amount *C*.

Table 11. Revised optimal pervious concrete mixes and test values

Mix	*W/C*	*C* (kg)	*W* (kg) Trial	*W* (kg) Field	*G* (kg) Trial	*G* (kg) Field	*U* (kg/m^3)	*Air* (%)	f_c (kgf/cm^2) 28th day	f_{c7}/f_{c28}	*R* (kgf/cm^2) 28th day	R_7/R_{28}	R_{28}/f_{c28}	*k* (cm/sec)	*n* (%)
Original	0.38	300	114	125	1488	1500	1976	18.4	194	0.82	31.4	0.67	0.16	0.083	36.7
Revised		308	117	128	1528	1540									
Original	0.40	350	140	150	1394	1406	1976	17.1	251	0.84	32.7	0.84	0.13	0.110	36.5
Revised		363	145	155	1445	1457									

7. Concluding remarks

The main grain size range of aggregate used in the tests is in #4~3/4″ (4.75~19.0 mm). The water-cement ratio *W/C*=0.36~0.44 of pervious concrete mixes of cement amount *C*= 300 and 350 kg was used to make fresh concrete samples to perform the drooping flow test and the unit weight and air content test. Then the specimens of pervious concrete were made to perform the compressive strength test, the flexural strength test, the permeability coefficient test and the porosity test in order to find the *W/C* of the maximum strength, which satisfies the requirements of permeability and retention. The mix of this *W/C* determined is the optimal economic effect mix.

The strength test results show that, the *W/C* values of the maximum strength are totally coincident either for *C*=300 kg or 350 kg, for the compressive strength f_c or the modulus of rupture *R*, for the 7th day or 28th day strengths. All maximum strengths are at *W/C*=0.38 for *C*=300 kg and at *W/C*=0.40 for *C*=350 kg. These results can be sure of that, the optimal mix is at *W/C*=0.38 for *C*=300 kg and at *W/C*=0.40 for *C*=350 kg. The optimal mix of *C*=300 kg gives strengths f_{c7} =159 kgf/cm^2 (15.6 MPa), f_{c28} =194 kgf/cm^2 (19.0 MPa), R_7 =22.6 kgf/cm^2 (2.2 MPa), R_{28} =31.4 kgf/cm^2 (3.1 MPa). Thus, it gives the strength ratios f_{c7}/f_{c28} = 0.82, R_7/R_{28} =0.67 and R_{28}/f_{c28}=0.16. The optimal mix of *C*=350 kg gives strengths f_{c7} =212 kgf/cm^2 (20.8 MPa), f_{c28} =251 kgf/cm^2 (24.6 MPa), R_7 =27.5 kgf/cm^2 (2.7 MPa), R_{28} =32.7 kgf/cm^2 (3.2 MPa). Thus, it gives the strength ratios f_{c7}/f_{c28} = 0.85, R_7/R_{28} =0.84 and R_{28}/f_{c28}=0.13.

The *W/C* values of the rising points at the drooping flow curves are totally coincident with those of the maximum strengths for *C*=300 kg and 350 kg. These results strengthen that, the drooping flow test is able to estimate the possible *W/C* range of the maximum strength of pervious concrete.

For *C*=300 kg, only the *W/C*=0.40 of obvious drop point at the air content line graph is not coincident with *W/C*=0.38, but still very close to it. For *C*=350 kg, only the *W/C*=0.42 of the permeability coefficient and the porosity tests is not totally coincident with *W/C*=0.40, but also very close to it. All the other test results are totally coincident with each other. Therefore, these test results provide evidences for the optimal mix determined above.

Only the *W/C*=0.40 of the air content test for *C*=300 kg is not totally coincident with *W/C*=0.38, but still very close to it. Only the *W/C*=0.42 of the permeability coefficient and the porosity tests for *C*=350 kg is not totally coincident with *W/C*=0.40, but also very close to it. Therefore, these test results provide evidences for the optimal mix determined above.

Since the measured air content is not the same with the assumed air content 20.5%, the original mix must be properly revised to produce 1 m^3 of concrete volume. The revised mixes are listed in the following table, and are the optimal mixes finally determined. From this table, it can be said that, the pervious concrete mixes of *C*=308 and 363 kg can optimally produce 194 kgf/cm^2 (19.0 MPa) and 251 kgf/cm^2 (24.6 MPa) of the 28th day compressive strength respectively. The mixes would give permeability coefficient *k*= 0.083 and 0.110 cm/sec and porosity *n*=36.7% and 36.5% respectively. These *k* values are much larger than the permeability coefficient requirement 10^{-2} cm/sec and the *n* values are much larger than the porosity requirement 15%. The results shown in a table 12 as below,

Table 12. Relationship between cement amount and f_c

		W (kg)		*G* kg)		*U*		f_c (kgf/cm^2)		*R* (kgf/cm^2)			*k*	
C (kg)	*W/C*	Trial	Field	Trial	Field	(kg/ m^3)	*Air* (%)	28th day	f_{c7}/f_{c28}	28th day	R_7/R_{28}	R_{28}/f_{c28}	(cm/ sec)	*n* (%)
308	0.38	117	128	1528	1540	1976	18.4	194	0.82	31.4	0.67	0.16	0.083	36.7
363	0.40	145	155	1445	1457	1976	17.1	251	0.84	32.7	0.84	0.13	0.110	36.5

From the above table, the relationship between cement amount *C* and the 28th day compressive strength may be established by interpolating as f_c =1.04*C*-125 kgf/cm^2 for *C*=300~370 kg. This formula can be used to estimate the compressive strength for a pervious concrete optimal mix of cement amount *C*.

8. References

[1] Caltrans 2014 *Pervious Pavement Design Guidance* (Sacramento: Office of Storm Water Management, Division of Design, Sacramento, California) chapter 1 pp 1-2

[2] PCC committee 2013 *Outline Specification for Public Construction* (Taipei: Public Construction Commission, Executive Yuan) v4.0 chapter 02794 p 4

[3] Ho M-G 2015 *Evaluation Manual for Green Building Material* (New Taipei: Taiwan Architecture & Building Center) p 77

[4] CNS 486 2001 Method of test for sieve analysis of fine and coarse aggregates (Taipei: Bureau of Standard, Metrology and Inspection, Ministry of Economic Affairs)

[5] CNS 488 1998 Method of test for density, relative density (specific gravity), and absorption of coarse aggregate (Taipei: Bureau of Standard, Metrology and Inspection, Ministry of Economic Affairs)

[6] Hu C-F 2015 *Study on MBOF Slag used for Pervious Concrete* (Pingtung: Department of Civil Engineering, National Pingtung University of Science and Technology) chapter 3 p13

[7] Pan S-J 2016 *Effect of Tamping Methods on the Strength and the Water Permeability of Pervious Concrete* (Pingtung: Department of Civil Engineering, National Pingtung University of Science and Technology) chapter 3 p13

[8] CNS 11151 1984 Method of test for unit weight, yield, and air content (gravimetric) of concrete (Taipei: Bureau of Standard, Metrology and Inspection, Ministry of Economic Affairs)

[9] CNS 1232 2002 Method of test for compressive strength of cylindrical concrete specimen (Taipei: Bureau of Standard, Metrology and Inspection, Ministry of Economic Affairs)

[10] CNS 1233 1984 Method of Test for flexural strength of concrete (using simple beam with third-point loading) (Taipei: Bureau of Standard, Metrology and Inspection, Ministry of Economic Affairs)

[11] CNS 14995 2006 Permeable concrete paving blocks (Taipei: Bureau of Standard, Metrology and Inspection, Ministry of Economic Affairs)

AUTHOR INDEX